HANDBOOK
OF
ELECTRONIC
COMMUNICATION

GARY M. MILLER

Dean, Engineering Technology
Monroe Community College
Rochester, New York 14623

HANDBOOK
OF
ELECTRONIC
COMMUNICATION

PRENTICE-HALL, INC., Englewood Cliffs, New Jersey 07632

Library of Congress Cataloging in Publication Data

MILLER, GARY M
　Handbook of electronic communication.

　Includes index.
　1. Telecommunication. 2. Electronics. I. Title.
TK5101.M496　　621.38　　78-11347
ISBN 0-13-377374-4

Editorial/production supervision and interior design
　　by Virginia Huebner
Cover design by Edsal Enterprises
Manufacturing buyer: Gordon Osbourne

©1979 by Prentice-Hall, Inc.
Englewood Cliffs, N.J. 07632

Printed in the United States of America
10　9　8　7　6　5　4　3　2　1

PRENTICE-HALL INTERNATIONAL, INC., *London*
PRENTICE-HALL OF AUSTRALIA PTY. LIMITED, *Sydney*
PRENTICE-HALL OF CANADA, LTD., *Toronto*
PRENTICE-HALL OF INDIA PRIVATE LIMITED, *New Delhi*
PRENTICE-HALL OF JAPAN, INC., *Tokyo*
PRENTICE-HALL OF SOUTHEAST ASIA PTE. LTD., *Singapore*
WHITEHALL BOOKS LIMITED, *Wellington, New Zealand*

CONTENTS

PREFACE

The field of electronic communications is most dynamic. This handbook provides practical and up-to-date coverage of the field for the following interest groups:

    I. The engineers and technicians presently working in the field of electronic communications.

    II. Specialists in the field of digital electronics who have a need to understand the basics of communications.

    III. Amateur radio enthusiasts with a desire to increase their knowledge of communications theory.

    IV. Students in virtually any of the many educational systems available for the study of electronics.

The questions and problems for each chapter at the end of the book are keyed to the pertinent chapter section. For example, problem 1-19-5 indicates chapter 1, the 19th problem, and section 1-5 contains the information for the problem solution. An asterisk before the question indicates a question taken from the

FCC "Study Guide and Reference Material for Commercial Radio Operator Examinations." The number following an FCC question indicates the element and question number. For example, the number 3.232 indicates that it is question 232 from element 3 of the FCC study guide. An S before the number (e.g. S3.232), indicates that the question is from the FCC supplement to the study guide. The FCC license is highly recommended to all in the field of electronics. Although certain jobs require specific types of FCC licenses, the prestige value of an FCC license is useful in getting a job, a promotion or a salary increase.

GARY M. MILLER

Rochester, New York

HANDBOOK
OF
ELECTRONIC
COMMUNICATION

1

NOISE AND BANDWIDTH

The function of a communication system is to transfer information from one point to another by means of a communication link. The first type of "information" that was electrically transferred was the human voice, in the form of code (the Morse code), which was then converted back to words at the receiving site. Human beings had a natural desire to communicate rapidly between distant points on the earth, and that initially was the major concern of these developments. As that goal became a reality, and with the evolution of new technology following the invention of the triode vacuum tube, new and less basic applications were also realized, such as entertainment, radar, television, and telemetry. The field of communications is still a highly dynamic one, with new semiconductors and advancing technology constantly making new equipment possible or allowing improvement in old systems. Communications was the basic origin of the electronics field, and no other major branch of electronics developed until the transistor made modern digital computers a reality. We now have two major subcategories in the field of electronics: communications and digital systems. As will be seen in Chapter 8, communications plays a major role in digital systems, and vice versa.

1-1 MODULATION

Basic to the field of communications is the concept of modulation. *Modulation* is the process of impressing information onto a high-frequency carrier for transmission. In essence, then, the transmission takes place at the high frequency (the carrier), which has been modified to "carry" the lower-frequency information. The low-frequency information is often termed the *intelligence signal,* or simply the *intelligence.* It follows, then, that once this information is received, the intelligence must be removed from the high-frequency carrier, a process known as *demodulation.* At this point, you may be thinking, why bother to go through this modulation/demodulation process? Why not just transmit the information directly? The problem is that the frequency of the human voice ranges from about 20 to 4000 Hz. If everyone transmitted those frequencies directly as radio waves, interference between them would cause them all to be ineffective. Another limitation of equal importance is that it is virtually impossible to transmit such low frequencies anyway, since the required antennas for efficient propagation would have to be miles in length.

The answer to these problems is modulation, which allows propagation of the low-frequency intelligence with a high-frequency carrier. The high-frequency carriers are chosen such that only one transmitter in an area operates at the same frequency to minimize interference, and that frequency is high enough such that efficient antenna sizes are manageable. There are three basic methods of impressing low-frequency information onto a higher-frequency carrier. Equation (1-1) is the mathematical representation of a sine wave which we shall assume to be the high-frequency carrier:

$$v = V_p \sin (\omega t + \phi) \tag{1-1}$$

where v = instantaneous value

$\quad V_p$ = peak value

$\quad \omega$ = angular velocity = $2\pi f$

$\quad \phi$ = phase angle

Any one of the last three terms could be varied in accordance with the low-frequency information signal so as to produce a modulated signal that contains the intelligence. If the amplitude term, V_p, is the parameter varied, it is termed *amplitude modulation* (AM). If the frequency is varied, it is termed *frequency modulation* (FM); varying the phase angle, ϕ, results in *phase modulation* (PM). In subsequent chapters, we shall study these systems in detail.

1-2 COMMUNICATION SYSTEMS

Communication systems are often categorized by the frequency of the carrier. Table 1-1 provides the names for the various ranges of frequencies in the radio spectrum. The extra-high-frequency range begins at the starting point of infrared

TABLE 1-1

Radio-Frequency Spectrum

Frequency	Designation	Abbreviation
30–300 Hz	Extremely low frequency	ELF
300–3000 Hz	Voice frequency	VF
3–30 kHz	Very low frequency	VLF
30–300 kHz	Low frequency	LF
300 kHz–3 MHz	Medium frequency	MF
3–30 MHz	High frequency	HF
30–300 MHz	Very high frequency	VHF
300 MHz–3 GHz	Ultra high frequency	UHF
3–30 GHz	Super high frequency	SHF
30–300 GHz	Extra high frequency	EHF

frequencies, but the infrareds extend considerably beyond 300 GHz (300 × 10^9 Hz). After the infrareds in the electromagnetic spectrum (of which the radio waves are a very small portion) come light waves, ultraviolet, X-rays, gamma rays, and cosmic rays.

A communication system can be very simple but can also assume very complex proportions. Figure 1-1 represents a simple system in block diagram form. Notice that the modulated stage accepts two inputs, the carrier and the information (intelligence) signal. It produces the modulated signal, which is subsequently amplified before transmission. Transmission of the modulated signal can take place by any one of three means: antennas, waveguides, or transmission lines. The receiving unit of the system then picks up the transmitted signal but must reamplify it to compensate for attenuation that occurred during its transmission. Once suitably amplified, it is fed to the demodulator (often referred to as the *detector*), where the information signal is extracted from the high-frequency carrier. The demodulated signal (intelligence) is then fed to the power amplifier. The signal is brought to a suitably high level by the power amplifier to drive a speaker or any other output transducer (load).

There are *two basic limitations* on the performance of a communication system: (1) electrical noise, and (2) the bandwidth of frequencies allocated for the transmitted signal. The rest of this chapter is devoted to these topics.

1-3 NOISE

Electrical noise may be defined as any undesired voltages or currents that ultimately end up appearing in the load of the communications receiver (usually a speaker). To the listener, this electrical noise often manifests itself as *static*. It may only be annoying, such as an occasional burst of static, or it may be continuous and of such amplitude that the desired information is obliterated.

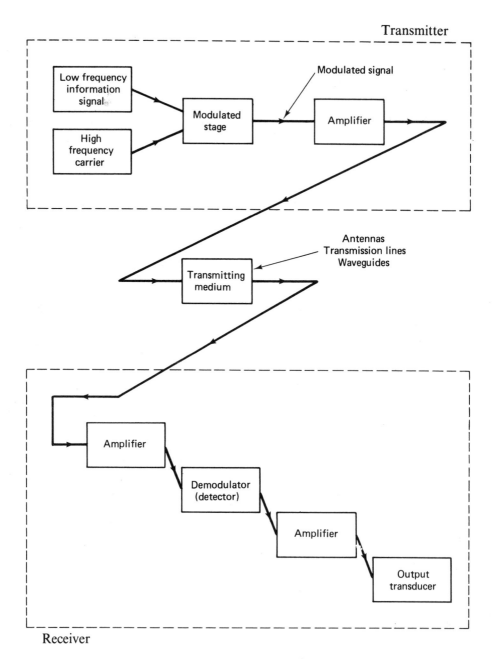

Fig. 1-1. Communication system block diagram.

Noise signals at their point of origin are generally very small—for instance, in the microvolt or millivolt level. You may be wondering, therefore, why they create so much trouble. Well, a communications receiver is a very sensitive instrument that is usually only given a very small signal at its input, which must be greatly amplified before it can drive a speaker. Consider the receiver block diagram shown in Fig. 1-1 to be representative for a standard FM radio (receiver). The first amplifier block, which forms the "front end" of the radio, is required to amplify a signal received from the radio's antenna, which is often less than 10 μV. It does not take a very large dose of undesired voltage (noise) to ruin reception. This is true even though the transmitted signal from the transmitter may be many thousands of watts, since when it reaches the receiver it is always severely attenuated. Therefore, if the desired signal received is of the same order of magnitude as the undesired noise signal, it is likely that the result will be unintelligible. This situation is made even worse by the fact that the receiver itself introduces noise in addition to the noise already present in the received signal.

The noise present in a received radio signal has been introduced in the transmitting medium and is termed *external noise*. The noise introduced by the radio receiver itself is termed *internal noise*. The important implications of noise considerations in the study of communication systems cannot be overemphasized.

External Noise

Man-made noise. The most troublesome form of external noise is usually of the man-made variety. It is often produced by spark-producing mechanisms such as engine ignition systems, fluorescent lights, and commutators in electric motors. This noise is actually *radiated* or transmitted from its generating source through the atmosphere in the same fashion that a transmitting antenna sends desirable electrical signals to a receiving antenna. If the man-made noise exists in the vicinity of the transmitted radio signal and contains some of the same frequencies, these two signals will "add" together. This is obviously an undesirable phenomenon. Man-made noise occurs randomly at frequencies up to approximately 500 MHz.

Another common source of man-made noise is contained on the power lines that supply the energy for most electronic systems. In this context, the ac ripple in a dc power supply output can be classified as noise (an unwanted electrical signal) and must be minimized in receivers that are accepting extremely small intelligence signals. The ac power lines contain surges of voltages caused by the switching on and off of heavy inductive loads such as motors. It is certainly ill-advised to operate sensitive electrical equipment in close proximity to such a load as an elevator! Since man-made noise is weakest in sparsely populated areas, this explains the locations of extremely sensitive communications equipment, such as satellite tracking stations, in desert-type locations.

Atmospheric noise. *Atmospheric noise* is caused by naturally occurring disturbances in the earth's atmosphere, lightning discharges being the most prominent contributors. Its frequency content is spread over the entire radio spectrum, but its intensity is inversely related to frequency. It is, therefore, most troublesome at the lower frequencies. It manifests itself in the static noise that you hear on standard AM radio receivers. It has the greatest intensity when a storm is in your vicinity but occurs in much greater quantity (but with less intensity) as a result of storms throughout the world. This is often apparent when listening to a distant station at night on an AM receiver. It is not a significant factor for frequencies exceeding about 20 MHz.

Space noise. The other form of external noise arrives from outer space and is therefore termed *space noise*. Space noise is pretty evenly divided in origin between the sun and all the other stars. That originating from our star (the sun) is termed *solar noise*. Solar noise is cyclical and reaches very annoying peaks every 11 years. These 11-year peaks are also cyclical, with the 1957 peak being the highest in recorded history. It was, therefore, selected by scientists around the world as the date for the International Geophysical Year, since this large increase in the sun's activity offered better opportunity to study its origin and characteristics.

All the other stars also generate this space noise, and their contribution is termed *cosmic noise*. Since they are much farther away than the sun, their individual effects are small, but they make up for this by their countless numbers and their additive effects. Space noise occurs at frequencies from about 8 MHz to over 1 GHz (10^9 Hz). While space noise contains energy at less than 8 MHz, these components are absorbed by the earth's ionosphere before they can reach the atmosphere. The ionosphere is a region above the atmosphere where free ions and electrons exist in sufficient quantity to have an appreciable effect on wave travel. It includes the area from about 60 miles up to several hundred miles above the earth.

Internal Noise

As previously stated, internal noise is that which is introduced by the receiver itself. Thus, the noise already present at the receiving antenna (external noise) has another component added to it before it reaches the receiver's output. The receiver's effective noise contribution is normally limited to its very first stage of amplification. It is there that the desired signal is at its lowest level, and noise injected at that point will be at its largest value in proportion to the intelligence signal. A glance at Fig. 1-2 should help clarify this point. Even though all following stages also introduce noise, those effects are usually negligible with respect to the very first stage because of the much higher signal level of the stages following the first one. Note that the noise injected between amplifiers 1 and 2 has not appreciably increased the noise on the desired signal, even though it is of the same magnitude as the noise injected into amplifier 1. For this reason, the very first receiver stage must be very carefully designed to have low noise characteristics, with the following stages being decreasingly important as the desired signal gets larger and larger.

Thermal noise. There are two basic types of noise generated by electronic circuits. The first is due to thermal interaction between the free electrons and

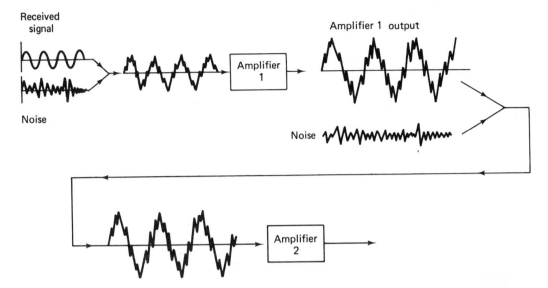

Fig. 1-2. Noise effect on a receiver's first and second amplifier stages.

vibrating ions in a conductor. Resistors are the major contributors, but noise exists within all other electrical devices. Thus, a resistor, all by itself, is constantly producing a voltage. This form of noise was first thoroughly studied by J. B. Johnson in 1928 and is often termed *Johnson noise*. Since it is dependent on temperature, it is also referred to as *thermal noise*. Its frequency content is spread equally throughout the usable spectrum, which leads to a third designator, *white noise* (from optics, where white light contains all frequencies or colors). The terms Johnson, thermal, and white noise may be used interchangeably. Johnson was able to show that the power of this generated noise is given by

$$P_{\text{noise}} = KT\,\Delta f \tag{1-2}$$

where K = Boltzmann's constant (1.38×10^{-23}) joule/°K

T = resistor temperature (°K)

Δf = bandwidth of frequencies that the subsequent amplifier is able to amplify

Since this noise power is directly proportional to the band of frequencies involved, it is advisable to limit the bandwidth of a receiver to the smallest usable value.

Since $P = E^2/R$, it is possible to rewrite Eq. (1-2) to determine the noise voltage generated by a resistor:

$$e_{\text{noise}} = \sqrt{4kT\,\Delta f R} \tag{1-3}$$

where e_{noise} = rms noise voltage generated

R = resistance (ohms)

The thermal noise associated with all nonresistor devices is a direct result of their inherent resistance and to a much lesser extent their composition. This applies to capacitors, inductors, and all electronic devices. Equation (1-3) applies to copper wirewound resistors, with all others exhibiting somewhat greater noise voltages. Thus, dissimilar resistors of equal value exhibit different noise levels, which gives rise to the term "low-noise resistor," which you may have heard before but not understood. Standard carbon resistors are the least expensive variety, but unfortunately they also tend to be the noisiest. Metal film resistors offer a good compromise in the cost/performance comparison and can be used in all but the most demanding low-noise designs. The ultimate noise performance (lowest noise generated, that is) is obtained with the most expensive and bulkiest variety—the wirewound resistor.

EXAMPLE 1-1

The amplifier in Fig. 1-3(a) has an effective bandwidth of 4 MHz, a voltage gain of 200, and operates at 27°C. Determine the rms noise and intelligence output signals assuming that external noise can be disregarded.

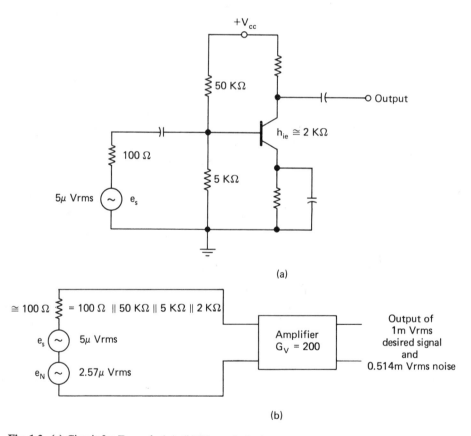

Fig. 1-3. (a) Circuit for Example 1-1. (b) It's equivalent.

Solution:

Its equivalent input resistance is $100\,\Omega\,\|\,50\,k\Omega\,\|\,5\,k\Omega\,\|\,h_{ie}$, where h_{ie} is the impedance looking into the base of Q_1. A typical value for a common emitter (CE) stage with an emitter bypass capacitor is $2\,k\Omega$, and thus the total resistance at the input of this amplifier seen by the generator is about $100\,\Omega\,\|\,50\,k\Omega\,\|\,5\,k\Omega\,\|\,2\,k\Omega$, or approximately $100\,\Omega$. This resistance can be termed the *equivalent noise resistance* and is the value which, when used with Eq. (1-3), can be used to predict the thermal noise generated at the amplifier's input. To convert °C to °K, simply add 273°, so that $°K = 27°C + 273°C = 300°K$. Therefore,

$$E_{\text{noise}} = \sqrt{4kT\,\Delta f R}$$
$$= \sqrt{4 \times 1.38 \times 10^{-23}\,\text{joule/}°K \times 300°K \times 4\,\text{MHz} \times 100\,\Omega}$$
$$= 2.57\,\mu V\ \text{rms}$$

After multiplying the intelligence signal e_s and noise signal by the voltage gain of 200, the output signal consists of a 1-mV rms signal and 0.514-mV rms noise. This is not normally an acceptable situation. The intelligence would probably be unintelligible!

Transistor noise. In Ex. 1-1, the noise introduced by the transistor, other than its thermal noise, was not considered. The major contributor of transistor noise is called *shot noise*. It is due to the discrete-particle nature of the current carriers in all semiconductors. These current carriers, even under dc conditions, are not moving in a continuous flow since the distance they travel is somewhat different for each carrier, because of their random motion. The name "shot noise" is derived from the fact that when driving a speaker, excessive shot noise sounds like a shower of lead shot falling on a metallic surface. Shot noise and thermal noise are additive. The equation for shot noise in a diode is:

$$i_{\text{noise}} = \sqrt{2qI_{dc}\,\Delta f} \tag{1-4}$$

where $i_{\text{noise}} =$ shot noise (rms amperes)

$\quad q =$ electron charge $(1.6 \times 10^{-19}\ \text{coulomb})$

$\quad I_{dc} =$ diode dc current (amperes)

$\quad \Delta f =$ bandwidth of frequencies involved (Hz)

Unfortunately, there is no valid formula to calculate its value for a complete transistor, where the sources of shot noise are the currents within the emitter-base and collector-base diodes. Hence, the device user must refer to the manufacturer's data sheet for an indication of shot noise characteristics. They often specify an *equivalent noise resistance*—that is, a resistance value that produces the same amount of noise as the device's shot noise when applied to the thermal noise of

Eq. (1-3). This resistance does not actually exist, it is only a convenient means of allowing the device user to calculate shot noise levels. Shot noise generally increases proportionally with dc bias currents except in MOSFETs, whose shot noise seems to be relatively independent of dc current levels.

Frequency noise effects. Two other little-understood forms of device noise occur at the opposite extremes of frequency. The low-frequency effect is called *excess noise* and occurs at frequencies below about 1 kHz. It is inversely proportional to frequency and directly proportional to temperature and dc current levels. It is thought to be associated with "traps" in the emitter depletion layer which capture and release carriers at different frequency rates but with energy levels that vary at an inverse rate with frequency. Excess noise is often referred to as *flicker noise*, or $1/f$ *noise*. It is present in both bipolar junction transistors (BJT) and field-effect transistors (FET). At the upper end of the frequency spectrum, device noise starts to increase rapidly in the vicinity of the device's high-frequency cutoff. These high- and low-frequency effects are relatively unimportant in the design of receivers, since the critical stages (the front end) will usually be working well above 1 kHz and hopefully below the device's high-frequency cutoff area. The low-frequency effects are, however, important to the design of low-level, low-frequency amplifiers encountered in certain instrument and biomedical applications.

The overall noise intensity-versus-frequency curves for semiconductor devices (and tubes) have a bathtub shape, as represented in Fig. 1-4. At low frequencies, the excess noise is dominant, while in the midrange, shot noise and thermal noise predominate and above that the high-frequency effects take over. Of course, tubes are now seldom used (for many reasons); fortunately, their semiconductor replacements offer better noise characteristics. Since semiconductors possess inherent resistances, they generate thermal noise in addition to shot noise, as indicated in Fig. 1-4. The noise characteristics provided in manufacturers' data sheets take into account both the shot and thermal effects. At the device's high-frequency cutoff, f_{hc}, the high-frequency effects take over and the noise increases rapidly.

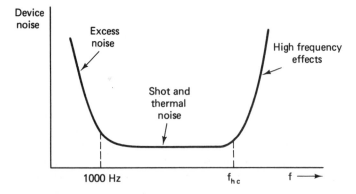

Fig. 1-4. Noise versus frequency curve.

following frequency components:

1. A dc level.
2. Components at each of the two original frequencies.
3. Components at the sum and difference frequencies of the two original frequencies.
4. Harmonics of the two original frequencies.

Figure 2-2 shows this process pictorially with the two sine waves, labeled f_c and f_i, to

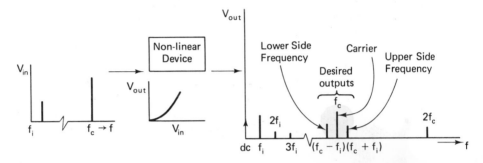

Fig. 2-2. Nonlinear mixing.

represent the carrier and intelligence frequencies, respectively. If all but the $f_c - f_i$, f_c, and $f_c + f_i$ components are removed, the three components that are left form an AM waveform. They are referred to as:

1. The *lower side frequency* $(f_c - f_i)$.
2. The carrier frequency (f_c).
3. The *upper side frequency* $(f_c + f_i)$.

AM Waveforms

Figure 2-3 shows the actual AM waveform under varying conditions of the intelligence signal. Note in Fig. 2-3(a) that the resultant AM waveform is basically a signal at the carrier frequency, but whose amplitude is changing at the same rate as the intelligence frequency is changing. Note that as the intelligence amplitude reaches a maximum positive value, the AM waveform has a maximum amplitude. The AM waveform reaches a minimum value when the intelligence amplitude is at a maximum negative value. In Fig. 2-3(b), the intelligence frequency remains the same, but its amplitude has been increased. The resulting AM waveform reacts by reaching a larger maximum value and smaller minimum value. In Fig. 2-3(c), the intelligence amplitude is reduced and its frequency has gone up. The resulting AM waveform, therefore, has reduced maximums and minimums, and the rate at

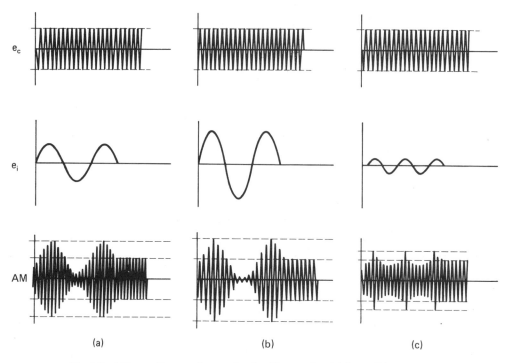

Fig. 2-3. AM waveform under varying intelligence signal (e_i) conditions.

which it swings between these extremes has increased to the same frequency as the intelligence signal.

It may now be correctly concluded that both the top and bottom envelopes of an AM waveform are replicas of the frequency and amplitude of the intelligence. However, the AM waveform does *not* include any component at the intelligence frequency. If a 1-MHz carrier were modulated by a 5-kHz intelligence signal, the AM waveform would include the following components:

$$1 \text{ MHz} + 5 \text{ kHz} = 10,005,000 \text{ Hz} \qquad \text{(upper side frequency)}$$
$$1 \text{ MHz} = 10,000,000 \text{ Hz} \qquad \text{(carrier frequency)}$$
$$1 \text{ MHz} - 5 \text{ kHz} = 995,000 \text{ Hz} \qquad \text{(lower side frequency)}$$

This process is shown in Fig. 2-4. Thus, even though the AM waveform has an envelope that is a replica of the intelligence signal, it *does not* contain a frequency component at the intelligence frequency.

The intelligence envelope is shown in the resultant waveform and results from connecting a line from each RF peak value to the next one for both the top and bottom halves of the AM waveform. The drawn-in envelope is not really a component of the waveform and would not be seen on an oscilloscope display. In

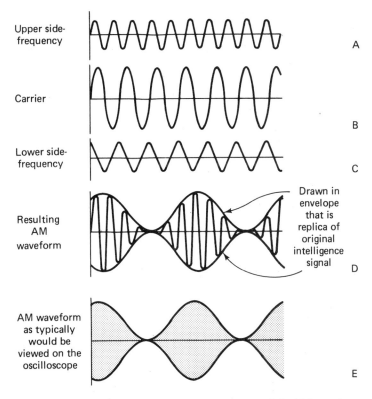

Fig. 2-4. Carrier and side-frequency components result in AM waveform.

addition, the top and bottom envelopes are *not* the upper and lower side frequencies, respectively. The envelopes are the result of combining a carrier with two smaller signals spaced in frequency equal amounts above and below the carrier frequency. The increase and decrease in the AM waveform's amplitude is caused by the frequency difference in the side frequencies, which allows them to alternately add to and subtract from the carrier amplitude, depending on their instantaneous phase relationships.

The AM waveform in Fig. 2-4(d) does not show the relative frequencies to scale. The ratio of f_c to the envelope frequency (which is also f_i) is 1 MHz to 5 kHz or 200:1. Thus, the fluctuating RF should show 200 cycles for every cycle of envelope variation. To do that in a sketch is not possible, and an oscilloscope display of this example, and most practical AM waveforms, results in a well-defined envelope but with so many RF variations that they appear as a blur in Fig. 2-4(e).

Modulation of a carrier with a pure sine-wave intelligence signal has thus far been shown. However, in most systems the intelligence is a rather complex waveform that contains many frequency components. For instance, the human voice contains components from roughly 200 Hz to 3 kHz and has a very erratic shape. If it were used to modulate the carrier, a whole *band* of side frequencies would be generated. The band of frequencies thus generated above the carrier is termed the *upper side band*, while those below the carrier are called the *lower side band*. This situation is illustrated in Fig. 2-5 for a 1-MHz carrier modulated by a whole band of frequencies, which range from 200 Hz up to 3 kHz. The upper side band is from 1,000,200 Hz to 1,003,000 Hz, and the lower side band ranges from 997,000 Hz to 999,800 Hz.

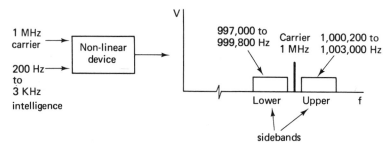

Fig. 2-5. Modulation by a band of intelligence frequencies.

EXAMPLE 2-1

A 1.4-MHz carrier is modulated by a music signal that has frequency components from 20 Hz to 15 kHz. Determine the range of frequencies generated for the upper and lower side bands.

Solution:

The upper side band is equal to the sum of carrier and intelligence frequencies. Therefore, the upper side band (usb) will include the frequencies from

$$1,400,000 \text{ Hz} + 20 \text{ Hz} = 1,400,020 \text{ Hz}$$

to

$$1,400,000 \text{ Hz} + 15,000 \text{ Hz} = 1,415,000 \text{ Hz}$$

The lower side band (lsb) will include the frequencies from

$$1,400,000 \text{ Hz} - 15,000 \text{ Hz} = 1,385,000 \text{ Hz}$$

to

$$1,400,000 \text{ Hz} - 20 \text{ Hz} = 1,399,980 \text{ Hz}$$

This result is shown in Fig. 2-6 with a frequency spectrum of the AM modulator's output.

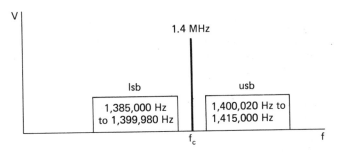

Fig. 2-6. Solution for Example 2-1.

2-2 PERCENTAGE MODULATION

In Sec. (2-1) it was determined that an increase in intelligence amplitude resulted in an AM signal with larger maximums and smaller minimums. It is helpful to have a mathematical relationship between the relative amplitude of the carrier and intelligence signals. The *percentage modulation* provides this, and it is a measure of the extent to which a carrier voltage is varied by the intelligence. The percentage modulation is also referred to as *modulation index* or *modulation factor*, and they are all rather loosely symbolized by *m*.

Figure 2-7 illustrates the two most common methods of determining the percentage modulation. Notice that when the intelligence signal is zero, the carrier

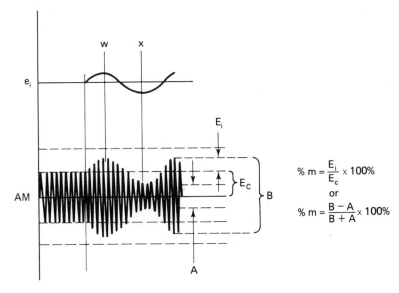

$$\% \, m = \frac{E_i}{E_c} \times 100\%$$

or

$$\% \, m = \frac{B - A}{B + A} \times 100\%$$

Fig. 2-7. Percentage modulation determination.

is unmodulated and has a peak amplitude labeled as E_C. When the intelligence reaches its first peak value (point W), the AM signal reaches a peak value labeled E_i above E_c. Percentage modulation is then given as

$$\%m = \frac{E_i}{E_c} \times 100\% \qquad (2\text{-}1)$$

or expressed simply by a ratio:

$$m = \frac{E_i}{E_c} \qquad (2\text{-}2)$$

The same result can be obtained by utilizing the maximum peak-to-peak value of the AM waveform (point w), which is shown as B, and the minimum peak-to-peak value (point x), which is A in the following equation:

$$\%m = \frac{B - A}{B + A} \times 100\% \qquad (2\text{-}3)$$

This method is usually more convenient in graphical (oscilloscope) solutions.

Overmodulation

If the AM waveform's minimum value A falls to zero as a result of an increase in the intelligence amplitude, the percentage modulation becomes

$$\%m = \frac{B - A}{B + A} \times 100\% = \frac{B - O}{B + O} \times 100\% = 100\%$$

This is the maximum possible degree of modulation. In this situation the carrier is being varied between zero and double its unmodulated value. Any further increase in the intelligence amplitude will cause a condition known as *overmodulation* to occur. If this does occur, the modulated carrier will go to more than double its unmodulated value but will fall to zero for an interval of time as shown in Fig. 2-8. This "gap" produces distortion termed *side-band splatter*, which results in the transmission of frequencies outside a station's normal allocated range. The FCC does not allow this condition, as it causes severe interference to other stations and causes a severe splattering sound to be heard at a receiver.

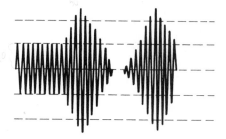

Fig. 2-8. Illustration of overmodulation.

Example 2-2

Determine the $\%m$ for the following conditions if the unmodulated carrier is 80 V peak to peak (p-p).

	Maximum p-p Carrier (V)	Minimum p-p Carrier (V)
A.	100	60
B.	125	35
C.	160	0
D.	180	0

Solution:

A. $\%m = \dfrac{B - A}{B + A} \times 100\%$ (2-3)

$= \dfrac{100 - 60}{100 + 60} \times 100\% = 25\%$

B. $\%m = \dfrac{125 - 35}{125 + 35} \times 100\% = 56.25\%$

C. $\%m = \dfrac{160 - 0}{160 + 0} \times 100\% = 100\%$

D. This is a case of overmodulation since the modulated carrier reaches a value more than twice its unmodulated value.

2-3 AM POWER RELATIONSHIPS

In the case where a carrier is modulated by a pure sine wave, it can be shown that at 100% modulation, the upper and lower side frequencies are one-half the amplitude of the carrier. In general,

$$E_{SF} = \frac{mE_c}{2} \qquad (2\text{-}4)$$

where E_{SF} = side-frequency amplitude

    m = modulation index

    E_c = carrier amplitude

In an AM transmission, the carrier amplitude and frequency always remains constant while the side bands are constantly changing in amplitude and frequency. Thus, the carrier contains no information since it never changes. However, it does contain the most power since its amplitude is always at least double (when m = 100%) the side band's amplitude. It is the side bands that contain the information.

EXAMPLE 2-3

Determine the maximum possible total side-band power if the carrier output is 1 kW and also calculate the total maximum transmitted power.

Solution:

Since

$$E_{SF} = \frac{mE_c}{2} \qquad (2\text{-}4)$$

it is obvious that the maximum side-band power occurs when $m = 1$ or 100%. At that percentage modulation, each side frequency has 1/2 the amplitude of the carrier and, since power is proportional to the square of voltage, each side band has 1/4 of the carrier power or 1/4 $\times$ 1 kW, or 250 W. Therefore, the total side-band power is 250 W $\times$ 2 = 500 W and the total transmitted power is 1 kW + 500 W, or 1.5 kW.

Importance of High-Percentage Modulation

It is important to use as high a percentage modulation as possible while ensuring that overmodulation does not occur. It is the side bands that contain the information and they have maximum power at 100% modulation. For example, if 50% modulation were used in Ex. 2-4, the side-band amplitudes are 1/4 the carrier amplitude and thus have $(1/4)^2$, or 1/16 the carrier power. Thus, total side-band power is now $1/16 \times 1$ kW $\times 2$, or 125 W. The actual transmitted intelligence is thus only 1/4 of the 500 W transmitted as full 100% modulation. These results are summarized in Table 2-1. Even though the total transmitted power has only fallen from 1.5 kW to 1.125 kW, the effective transmission has only 1/4 the strength at 50% modulation as compared to 100%. Because of these considerations, most AM transmitters attempt to maintain between 90 and 95% modulation as a compromise between efficiency and the chance of drifting into overmodulation.

TABLE 2-1

Modulation Index (m)	Carrier Power	Power in One Side Band	Total Side Band Power	Total Transmitted Power (P_t)
0	1 kW	250 W	500 W	1.5 kW
0.5	1 kW	62.5 W	125 W	1.125 kW

A valuable relationship for many AM calculations is

$$P_t = P_c\left(1 + \frac{m^2}{2}\right) \qquad (2\text{-}5)$$

where P_t = total transmitted power of side bands and carrier

P_c = carrier power

m = modulation index

Equation (2-5) can be manipulated to utilize currents instead of powers. This is a useful relationship since they are usually the most easily measured quantities of a transmitter's output into its antenna.

$$I_t = I_c\sqrt{1 + \frac{m^2}{2}}$$

where I_t = total transmitted current

I_c = carrier current

m = modulation index

EXAMPLE 2-4

A 500-W carrier is to be modulated to a 90% level. Determine the total transmitted power.

Solution:

$$P_t = P_c\left(1 + \frac{m^2}{2}\right) \qquad (2\text{-}5)$$

$$= 500 \text{ W}\left(1 + \frac{0.9^2}{2}\right)$$

$$= 702.5 \text{ W}$$

EXAMPLE 2-5

An AM broadcast station operates at its maximum allowed total output of 50 kW and at 95% modulation. How much of its transmitted power is intelligence?

Solution:

$$P_t = P_c\left(1 + \frac{m^2}{2}\right)$$

$$50 \text{ kW} = P_c\left(1 + \frac{0.95^2}{2}\right)$$

$$P_c = \frac{50 \text{ kW}}{1 + (0.95^2/2)}$$

$$= 34.5 \text{ kW}$$

Therefore, the total intelligence signal is

$$P_i = P_t - P_c = 50 \text{ kW} - 34.5 \text{ kW}$$

$$= 15.5 \text{ kW}$$

EXAMPLE 2-6

The antenna current of an AM transmitter is 12 A when unmodulated but increases to 13 A when modulated. Calculate %m.

Solution:

$$I_t = I_c\sqrt{1 + \frac{m^2}{2}} \qquad (2\text{-}6)$$

$$13 \text{ A} = 12 \text{ A}\sqrt{1 + \frac{m^2}{2}}$$

$$1 + \frac{m^2}{2} = \left(\frac{13}{12}\right)^2$$

$$m^2 = 2\left[\left(\frac{13}{12}\right)^2 - 1\right]$$

$$= 0.34$$

$$m = 0.58$$

$$\%m = 0.58 \times 100\% = 58\%$$

EXAMPLE 2-7

An intelligence signal is amplified by a 70% efficient amplifier before being combined with a 10-kW carrier to generate the AM signal. If it is desired to operate at 100% modulation, what is the dc input power to the final intelligence amplifier?

Solution:

You may recall that the efficiency of an amplifier is the ratio of ac output power to dc input power. To fully modulate a 10-kW carrier requires 5 kW of intelligence. Therefore, to provide 5 kW of sideband (intelligence) power through a 70% efficient amplifier requires a dc input of

$$\frac{5\,\text{kW}}{0.70} = 7.14\,\text{kW}$$

2-4 *CIRCUITS FOR AM GENERATION*

Amplitude modulation is generated by combining carrier and intelligence frequencies through a nonlinear device. A diode has a nonlinear area, but they are not often used because, being passive devices, they offer no gain. Transistors offer nonlinear operation (if properly biased) and provide amplification, thus making them ideal for this application. Figure 2-9(a) shows an input/output relationship for a typical BJT. Notice that at both low and high values of current, nonlinear areas exist. Between these two extremes is the linear area that should be used for normal amplification. The nonlinear areas must be used to generate AM.

Figure 2-9(b) shows a very simple transistor modulator. It operates with no base bias and thus depends on the positive peaks of e_c and e_i to bias it into the first nonlinear area shown in Fig. 2-9(a). Proper adjustment of the levels of e_c and e_i is necessary for good operation. Their levels must be low to stay in the first nonlinear area, and the intelligence power must be one-half the carrier power (or less) for 100% modulation (or less). In the collector circuit a parallel resonant circuit, tuned to the carrier frequency, is used to tune into the three desired frequencies—the upper and lower side bands and the carrier. The resonant circuit presents a

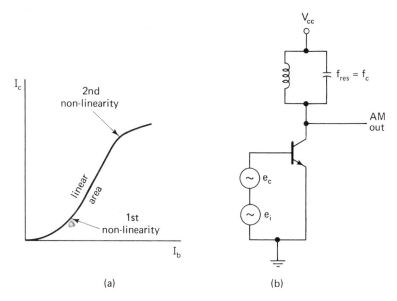

(a) (b)

Fig. 2-9. Simple transistor modulator.

high impedance to the carrier (and any other close frequencies such as the side bands) and thus allows a high output to those components, but its very low impedance to all other frequencies effectively shorts them out. Recall that the mixing of two frequencies through a nonlinear device generates more than just the desired AM components, as illustrated in Fig. 2-2. The tuned circuit then "sorts" out the three desired AM components and serves to provide good sinusoidal components by the flywheel effect.

In practice, amplitude modulation can be obtained in a number of ways. For descriptive purposes, the point of intelligence injection is utilized. For example, in Fig. 2-9(b) the intelligence is injected into the base, hence is termed *base modulation*. *Collector* and *emitter modulation* are also used. In previous years, when vacuum tubes were widely used, the most common form was *plate modulation*, but *grid*, *cathode*, and (for pentodes) *suppressor-grid* and *screen-grid* modulation schemes were also utilized.

High- and Low-Level Modulation

Another common designator for modulators involves whether or not the intelligence is injected at the last possible place or not. For example, the plate-modulated circuit shown in Fig. 2-10 has the intelligence added at the last possible point before the transmitting antenna and is termed a *high-level* modulation scheme. If the intelligence was injected at any previous point, such as the base, emitter, or even a previous stage, it would be termed *low-level* modulation. The

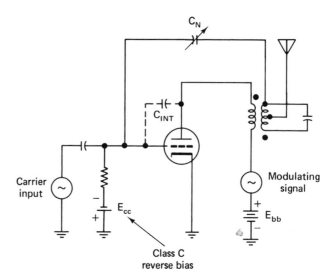

Fig. 2-10. Plate-modulated class C amplifier.

designer's choice between high- and low-level systems is made largely on the basis of the required power output. For high-power applications such as standard radio broadcasting, where outputs are measured in terms of kilowatts instead of watts, high-level modulation is the most economical approach. Recall that class C bias (device conduction for less than 180°) allows for the highest possible efficiency. It is nearly 90% efficient as compared to about 70% for the next best configuration, which is a class B amplifier. However, class C amplification cannot be used for reproduction of the AM signal, and hence large amounts of intelligence power must be injected at the final output to provide a high percentage modulation.

In summary, then, high-level modulation requires larger intelligence power to produce modulation but allows extremely efficient amplification of the higher-powered carrier. Low-level schemes allow low-powered intelligence signals to be used, but all subsequent output stages must use less efficient linear (not class C) configurations. Low-level systems usually offer the most economical approach for low-power transmitters.

Neutralization

One of the last remaining applications where tubes offer advantages over solid-state devices is in radio transmitters, where kilowatts of output power are required at high frequencies. Thus, the general configuration shown in Fig. 2-10 is still being utilized. Note the variable capacitor, C_N, connected from the plate tank circuit back to the grid. It is termed the *neutralizing* capacitor. It provides a path for the return of a signal that is 180° out of phase with the signal returned from plate to grid via the internal interelectrode capacitance of the tube. C_N is adjusted to cancel the internally fed-back signal to reduce the tendency of self-oscillation. The transformer in the plate is made to introduce a 180° phase shift by appropriate wiring. The process of *neutralization* is also used in transistor modulators and RF amplifiers.

Transistor High-Level Modulator

Figure 2-11 shows a transistorized class C, high-level modulation scheme. Class C operation provides an abrupt nonlinearity when the device switches "on" and "off," which allows for the generation of sum and difference frequencies. This is in contrast to the use of the gradual nonlinearities offered by a transistor at high and low levels of class A bias, as previously shown in Fig. 2-9(a). Generally, the operating point is established so as to allow half the maximum ac output voltage to be supplied at the collector when the intelligence signal is zero. The V_{bb} supply provides a reverse bias for Q_1 so that it conducts on only the positive peak of the input carrier signal. This, by definition, is class C bias, since Q_1 conducts for less than 180° per cycle. The tank circuit in Q_1's collector is tuned to resonate at f_c, and thus the full carrier sine wave is reconstructed there by the flywheel effect at the extremely high efficiency afforded by class C operation.

The intelligence signal for the collector modulator of Fig. 2-11 is added

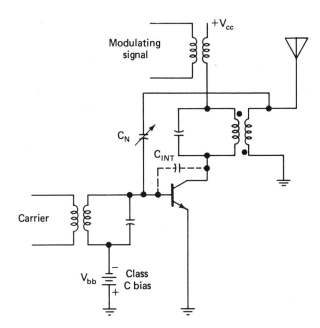

Fig. 2-11. Collector modulator.

directly in series with the collector supply voltage. The net effect of the intelligence signal is to vary the energy available to the tank circuit each time Q_1 conducts on the positive peaks of carrier input. This causes the output to reach a maximum value when the intelligence is at its peak positive value and a minimum value when the intelligence is at its peak negative value. Since the circuit is biased so as to provide one-half of the maximum possible carrier output when the intelligence is zero, theoretically an intelligence signal level exists where the carrier will swing between twice its static value and zero. This is a fully modulated (100% modulation) AM waveform. In practice, however, the collector modulator cannot achieve 100% modulation, because the transistor's knee in its characteristic curve changes at the intelligence frequency rate. This limits the region over which the collector voltage can vary, and slight collector modulation of the preceding stage is necessary to allow the high modulation indexes that are usually desirable. This is sometimes not a necessary measure in the tube-type high-level modulators.

Figure 2-12(a) shows an intelligence signal for a collector modulator, and Fig. 2-12(b) shows its effect on the collector supply voltage. In Fig. 2-12(c), the resulting collector current variations that are in step with the available supply voltages are shown, and Fig. 2-12(d) shows the collector voltage produced by the flywheel effect of the tank circuit as a result of the varying current peaks that are flowing through the tank.

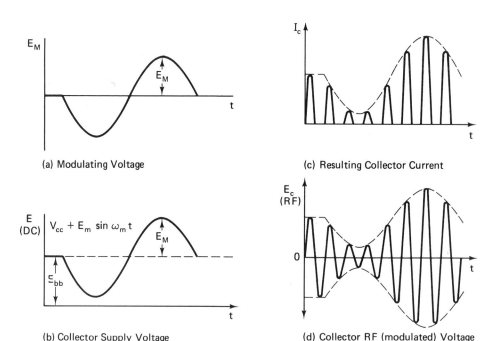

(a) Modulating Voltage

(b) Collector Supply Voltage

(c) Resulting Collector Current

(d) Collector RF (modulated) Voltage

Fig. 2-12. Collector modulator waveforms.

Linear-Integrated-Circuit Modulators

The process of generating high-quality AM signals economically is greatly simplified by the availability of low-cost specialty linear integrated circuits (LIC). This is especially true for low-power systems, where low-level modulation schemes are attractive. As an example, the RCA CA3080 operational transconductance amplifier (OTA) can be used to provide AM with an absolute minimum of design considerations. The OTA is similar to conventional operational amplifiers inasmuch as they employ the usual differential input terminals, but its output is best described in terms of the output current, rather than voltage, that it can supply. In addition, it contains an additional control terminal which enhances its flexibility for use in a variety of applications, including AM generation.

Figure 2-13(a) shows the CA3080 connected as an amplitude modulator. The gain of the OTA to the input carrier signal is controlled by variation of the amplifier-bias current (I_{ABC}), because the OTA transconductance (and hence gain) is directly proportional to this current. The level of the unmodulated carrier output is determined by the quiescent I_{ABC} current which is set by the value of R_M. The 100-kΩ potentiometer is adjusted to set the output voltage symmetrically about zero, thus nulling the effects of amplifier input offset voltage. Figure 2-13(b) shows the following:

> *Top trace:* the original intelligence signal superimposed on the upper AM envelope which gives an indication of the high quality of this AM generator.
> *Center trace:* the AM output.
> *Lower trace:* the AM output with the scope's vertical sensitivity greatly expanded to show the ability to provide high degrees of modulation (99% in this case) with a high degree of quality.

Another LIC modulator is shown in Fig. 2-14(a). This circuit uses an HA-2735 programmable dual op amp—half of it to generate the carrier frequency (A_1) and the other half as the AM generator (A_2).

In the HA-2735 op amp, the set current (pin 1 for A_L and pin 13 for A_2) controls the frequency response and gain of each amplifier. This "programmable" function is unnecessary for the oscillator circuit, and thus $I_{set\,1}$ is fixed by the 147-kΩ resistor. Carriers up to about 2 MHz can be generated with A_1.

Amplifier A_1 operates as a Wien-bridge oscillator. Amplitude control of the oscillator is achieved with 1N914 clamping diodes in the feedback network. If the output voltage tends to increase, the diodes offer more conductance, which lowers the gain. Resistor R_1 is adjusted to minimize distortion from nonlinear diode action. With the components in Fig. 2-14(a), the carrier frequency is approximately 700 kHz. This frequency can be changed by selection of different *RC* combinations in the Wien-bridge feedback circuit (1 kΩ and 120 pF).

Amplifier A_2's open-loop response is controlled by the modulating voltage applied to R_2. The percentage of modulation is directly proportional to the modu-

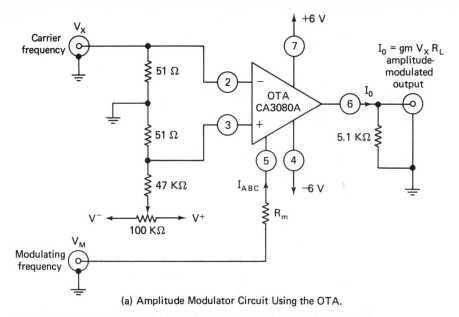

(a) Amplitude Modulator Circuit Using the OTA.

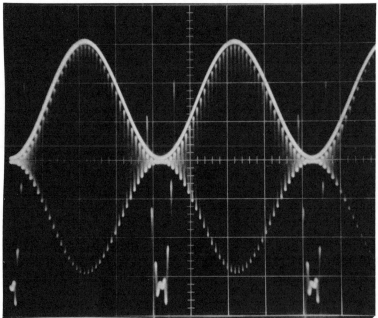

(b) Top Trace: Modulation frequency input
     ≈ 20 Volts P-P & 50µsec/DIV
   Center Trace: Amplitude modulate output
     500 mV/DIV & 50µsec/DIV
   Bottom Trace: Expanded output to show
     depth of modulation 20mV/DIV
     & 50µsec/DIV

Fig. 2-13. Linear Integrated Circuit (LIC) modulator and resulting waveforms. (Courtesy of RCA Solid State Division)

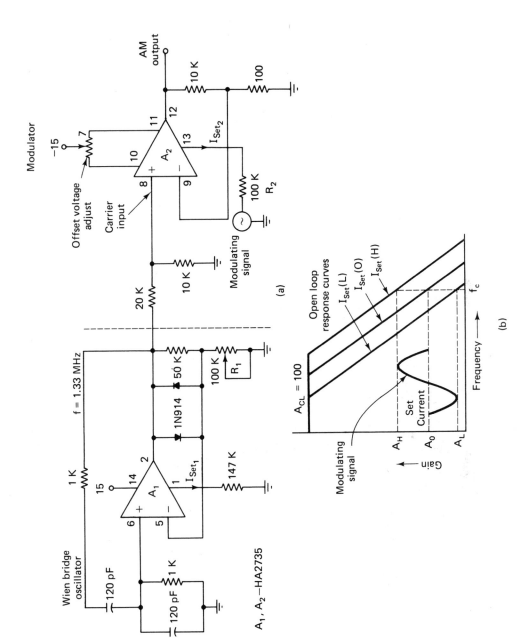

Fig. 2-14. LIC modulator.

43

lating voltage, but the output envelope is 180 degrees out of phase with the voltage. When sinusoidal modulation is applied to R_2, the circuit gain varies from a maximum, A_H, to a minimum, A_L, as A_2's frequency response to the carrier frequency, f_c, is modulated by the set current [Fig. 2-14(b)]. This results in a very distortion-free AM output at pin 12 of A_2. This circuit makes an ideal test generator for troubleshooting AM systems with carrier frequencies to 2 MHz. If a crystal oscillator were used instead of the Wien-bridge circuit, a high-quality AM transmitter can be fabricated with this AM generator.

2-5 AM TRANSMITTER SYSTEMS

Section (2-4) dealt with specific circuits to generate AM. Those circuits are only one element of a transmitting system. It is important to obtain a good understanding of a complete transmitting unit, and that is the goal of this section.

Figure 2-15 provides block diagrams of simple high- and low-level AM transmitters. The oscillator that generates the carrier signal will invariably be crystal-controlled to maintain the high accuracy required by the FCC. It is followed by the "buffer" amplifier, which provides a high impedance load for the oscillator to minimize drift. It also provides enough gain to sufficiently drive the modulated amplifier. Thus, the buffer amplifier could be a single stage, or however many stages are necessary to drive the following stage, the modulated amplifier.

The intelligence amplifier receives its signal from the input transducer (often a microphone) and contains whatever stages of intelligence amplification are required except for the last one. The last stage of intelligence amplification is called the *modulator*, and its output is mixed in the following stage with the carrier to generate the AM signal. The stage that generates this signal is termed the *modulated*

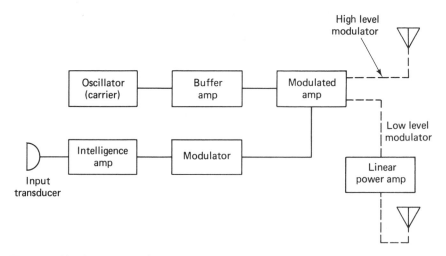

Fig. 2-15. Simple AM transmitter block diagram.

amplifier. This is also the output stage for high-level systems, but low-level systems have whatever number (one or more) of output stages are required. Recall that these stages are now amplifying the AM signal and must, therefore, be linear (class A or B), as opposed to the more efficient but nonlinear class C amplifier that can be used as an output stage in high-level schemes.

Citizens'-Band Transmitter

Figure 2-16 provides a typical AM transmitter configuration for use on the 27 MHz class D citizens' band. It is taken from the Motorola Semiconductor Products, Inc., Applications Note AN596. It is designed for 13.6-V dc operation, which is the typical voltage level in standard 12-V automotive electrical systems. It employs low-cost plastic transistors and features a novel high-level collector modulation method using two diodes and a double-pi output network for matching to the antenna impedance and harmonic suppression.

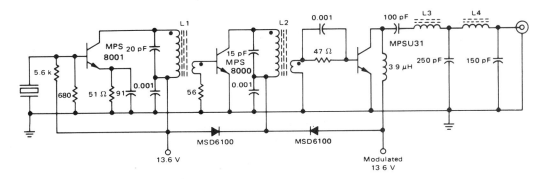

Fig. 2-16. Class D Citizen's Band transmitter. (Courtesy of Motorola Semiconductor Products, Inc.)

The first stage uses a MPS 8001 transistor in a common-emitter Colpitts oscillator configuration. Notice the split capacitor and L1 primary combination that is the heart of the Colpitts oscillator and the 27-MHz crystal, which provides excellent frequency stability with respect to temperature and supply voltage variations (well within the 0.005% allowance by FCC regulations for this band). This RF oscillator delivers about 100 mW of 27-MHz carrier power through the L1 coupling transformer into the buffer (alternately termed the *driver*) amplifier, which uses a MPS 8000 transistor in a common-emitter configuration. Information that allows fabrication of the coils used in this transmitter is provided in Fig. 2-17. The use of coils such as these is a necessity in transmitters and allows for required impedance transformations, interstage coupling, and "tuning" into desired frequencies when combined with the appropriate capacitance to form electrical resonance.

The buffer drives about 350 mW into the modulated amplifier, which can also

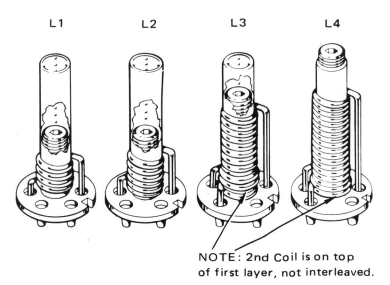

NOTE: 2nd Coil is on top
of first layer, not interleaved.

Fig. 2-17. Coil description for transmitter shown in Fig. 2-16: Conventional transformer coupling is employed between the oscillator and driver stages (L1) and the driver and final stages (L2). To obtain good harmonic suppression, a double-pi matching network consisting of L3 and L4 is utilized to couple the output to the antenna. All coils are wound on standard 1/4″ coil forms with #22 AWG wire. Carbonyl "J" 1/4″ × 3/8″ long cores are used in all coils. Secondaries are overwound on the bottom of the primary winding. The "cold" end of both windings is the start (bottom) and both windings are wound in the same direction. L1—Primary: 12 turns (closewound); Secondary: 2 turns overwound on bottom of primary winding. L2—Primary: 18 turns (closewound); Secondary: 2 turns overwound on bottom of primary winding. L3—7 turns (closewound); L4—5 turns (closewound). (Courtesy of Motorola Semiconductor Products, Inc.)

be termed the final transistor. It uses a MPSU31 RF power transistor and can subsequently drive 3.5 W of AM signal into the antenna. This system uses high-level collector modulation on the MPSU31 final transistor, but to obtain a high modulation percentage it is necessary to collector-modulate the previous MPS8000 transistor. This is accomplished with the aid of the MSD6100 dual diode shown in Fig. 2-16. The point labeled "modulated 13.6 V" is the injection point for the intelligence signal riding on the dc supply level of 13.6 V dc. A complete system block diagram for this transmitter is shown in Fig. 2-18. To obtain 100% modulation requires about 2.5 W of intelligence power. Thus, the audio amplification blocks between the microphone and the coupling transformer could easily be accomplished by a single low-cost LIC audio power amp (shown in dotted lines) capable of 2.5 W of output. The audio output is combined with the dc by the coupling transformer, as is required to modulate the 27-MHz carrier.

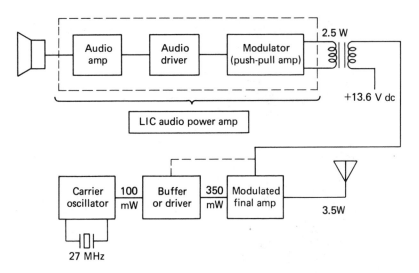

Fig. 2-18. Citizen's Band transmitter block diagram.

Antenna Coupler

Once the 3.5 W of AM signal is obtained at the final stage (MPSU31), it is necessary to "couple" this signal into the antenna. The coupling network for this system is comprised of L_3, L_4, and the 250-pF and 150-pF capacitors in Fig. 2-16. This filter configuration is termed a *double-pi network*. To obtain maximum power transfer to the antenna, it is necessary that the transmitter's output impedance be properly matched to the antenna's input impedance. This means equality in the case of a resistive antenna or the complex conjugate in the case of a reactive antenna input. If the transmitter were required to operate at a number of different carrier frequencies, the coupling circuit is usually made variable to obtain maximum transmitted power at each frequency. Coupling circuits are also required to perform some filtering action (to eliminate unwanted frequency components), in addition to their efficient energy transfer function. Conversely, a filter invariably performs a coupling function, and hence the two terms (filter and coupler) are really interchangeable, with what they are called generally governed by the function considered of major importance.

The double-pi network used in the citizen's band transmitter is very effective in suppressing (i.e., filtering out) the second and third harmonics, which would otherwise interfere with communications at 2×27 MHz and 3×27 MHz. It typically offers 37-dB second harmonic suppression and 55-dB third harmonic suppression. The capacitors and inductors in the double-pi network are resonant so as to allow frequencies in the 27-MHz region (the carrier and side bands) to pass, but all other frequencies are severely attenuated. The ratio of the values of the two capacitors determines what part of the total impedance across L_4 is coupled to the antenna, and the value of the 150-pF capacitor has a direct effect on the output impedance.

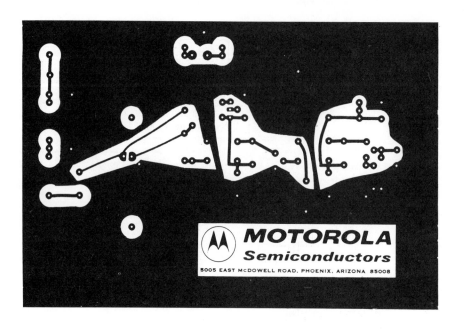

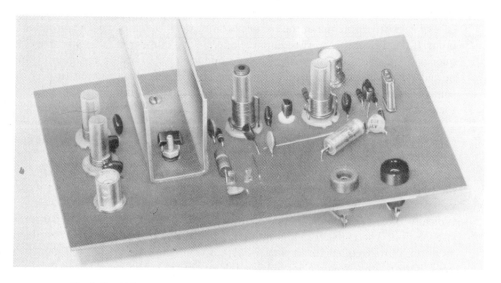

Fig. 2-19. Citizen's Band transmitter PC board layout and complete assembly pictorial. (Courtesy of Motorola Semiconductor Products, Inc.)

48

Transmitter Fabrication and Tuning

The fabrication of high-frequency circuits is much less straightforward than for low frequencies. The minimal inductance of a simple conductor or capacitance between two adjacent conductors can play havoc at high frequencies. Common sense and experimentation generally yield a suitable configuration. The information contained in Fig. 2-19 provides a suggested printed-circuit-board layout and component mounting photograph for the high-frequency sections (shown schematically in Fig. 2-16).

After assembly it is necessary to go through a "tune-up" procedure to get the transmitter on the air. Initially, L_1's variable core must be adjusted to get the oscillator to oscillate. This is necessary to get its inductance precisely adjusted so that, in association with its shunt capacitance, it will resonate at the precise 27-MHz resonant frequency of the crystal. The tune-up procedure starts by adjusting the cores of all four coils one-half turn out of the windings. Then tune L_1 clockwise until the oscillator starts and continue for one additional turn. Ensure that the oscillator starts every time by turning the dc on and off (a process termed *keying*) a number of times. If it does not reliably start every time, turn L_1 clockwise 1/4 turn at a time until it does. Then tune the other coils in order, with the antenna connected, for maximum power output. Apply nearly 100% sine-wave intelligence modulation and retune L_2, L_3, and L_4 once again for maximum power output while observing the output on an oscilloscope to ensure that overmodulation and/or distortion do not occur.

2-6 MONOLITHIC LIC TRANSMITTERS

The use of monolithic LIC technology is beginning to be applied to the field of communication transmitters. Instead of the discrete circuitry utilized in the previously discussed transmitter, a large majority of it can be replaced by a single LIC chip. The advantages offered by the approach include:

1. Cost reduction.
2. Improved reliability.
3. Size and weight reductions.

These advantages will make new applications available, such as the mass use of personal short-range two-way radios, remote biomedical patient monitoring systems, and compact security alarms.

A monolithic transmitter offered by Lithic Systems, Inc., is shown in Fig. 2-20. The block diagram of the chip shows that the functions of a complete transmitter are indeed available. The schematic at (b) shows the external discrete components required to make a complete AM transmitter system. Obviously, a very compact transmitter results. The power output of 50 mW is adequate for short-range communications of 1 mile or less, but certainly this output could be fed into a power amplifier stage if necessary. The unit can be operated at carrier frequencies up to about 100 MHz and thus a great range of the radio spectrum can be accom-

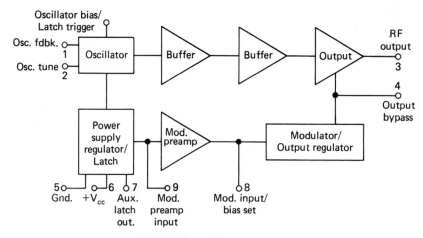

(a) Chip Block Diagram

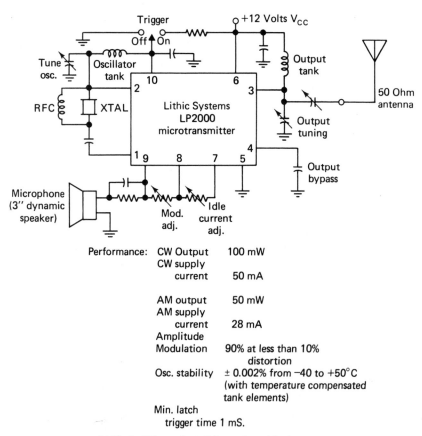

Performance: CW Output   100 mW
              CW supply
                current   50 mA

              AM output   50 mW
              AM supply
                current   28 mA
              Amplitude
              Modulation   90% at less than 10%
                                distortion
              Osc. stability ± 0.002% from −40 to +50°C
                              (with temperature compensated
                              tank elements)

              Min. latch
                trigger time 1 mS.

(b) Typical Transmitter Schematic and Performance

Fig. 2-20. A monolithic transmitter. (Courtesy of Lithic Systems, Inc.)

plished by simply changing the oscillator and output tank elements and the crystal. The advantages of complex circuitry arrangements within the compact area of a LIC has eliminated the need for the bulky *LC* tank couplings between the oscillator, buffers, and output stage.

Internal power supply regulation is provided to increase crystal oscillator stability. The regulator is normally "off" when first connected to the external supply, and contains a latch function, which may be externally triggered in the "transmit" mode or connected to turn on when power is applied. The latch causes the entire circuit to draw no power in the "off" condition, which allows it to be permanently connected across standby batteries in alarm applications. The unit will, therefore, draw power only when triggered. A latch output, pin 7, is available to switch on other circuitry, such as sources of modulating information or subsequent power amplifiers, when triggered "on".

2-7 AM STANDARD-BROADCAST-BAND TRANSMITTER

In Sec. (2-6) a specific transmitter used to communicate with a remote location was examined. Literally millions of similar transmitters (and, therefore, receivers) exist, utilizing all forms of modulation and their variations at every conceivable carrier frequency. Their applications are just as diverse with a partial list of users shown below:

1. Police departments.
2. Fire departments.
3. Taxi companies.
4. Appliance repair shops.
5. Hobbyists.
6. Military organizations.
7. Aircraft industry.
8. Radio-telephone industry.
9. Industrial security.
10. Entertainment and information (standard broadcast).

Obviously, the list could go on, but this gives you an idea of the magnitude of the application of radio communications to everyday life. Transmitter output powers from the milliwatt region up to several hundred watts are generally suitable for all but the last application on the list. The standard broadcast bands encompass 540 to 1600 kHz for the AM band; 88 to 108 MHz for FM; and 54 to 88, 176 to 216, and 470 to 890 MHz for television. To reach as many listeners as possible, transmitter powers up to the maximum allowable FCC limit of 50 kW are utilized. The broadcast band for AM allows 10 kHz per channel, which means that the maximum intelligence frequency is 5 kHz.

Broadcast transmitters involve special design considerations because of the higher powers involved and also because the ultimate in signal quality and system reliability is required. The reader can gain much insight into typical broadcast transmitters by studying the catalog information included in Fig. 2-21 for a Harris

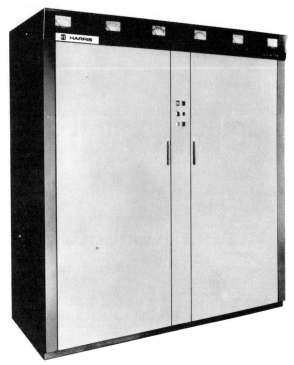

MODEL BC-10H

The most outstanding 10,000-watt AM transmitter available today, Harris' BC-10H is capable of providing the maximum positive modulation peaks allowed by the FCC (125%), with plenty of reserve for great reliability. Excellent signal quality and low operating costs are other proven features that help make the BC-10H number one in its power range.

SOLID-STATE CIRCUITRY. The BC-10H uses transistors in all circuits except the RF driver, power amplifier and modulator to provide a richer, fuller sound for the listener, and increased reliability for the broadcaster.

LOW TUBE COST. Ceramic type 3CX2500F3 triode tubes are used in the power amplifier and modulators, and a type 4-400A tetrode is used as the RF driver. This combination provides the lowest cost tube complement of any 10 kW AM transmitter on the market. All tubes are operated well below their maximum ratings for long tube life. Typically, 16,000 to 18,000 hours have been reported by many BC-10H users.

RF SECTION. Two transistor oscillators are instantly switchable, and oscillator output is amplified to provide the proper signal level for the driver, a 4-400A tetrode, which is modulated to improve the overall transmitter performance. The 4-400A drives two 3CX2500F3 power amplifiers which are high level plate modulated. These air-cooled power amplifiers have an efficiency as high as 90%, delivering full power through a full Pi-Tee network. The RF output capability of the BC-10H, 10,800 watts, easily accommodates complicated multi-tower phasors.

AUDIO SECTION. A solid-state audio driver provides full audio power direct to the grids of the two 3CX2500F3 modulator tubes. This combination is capable of more than 125% positive peak modulation if not limited by external amplifiers. Inverse feedback, and an advanced design low-leakage reactance modulation transformer/reactor group, results in signal quality of the highest fidelity. The modulation transformer is oil (Askarel) filled.

Fig. 2-21. Gates 10 kW broadcast transmitter. (Courtesy of Harris Corp.)

52

EFFICIENT COOLING. Individual Rotron blowers in the RF and modulator stages, and a specially designed air exhaust, allow only a limited amount of direct heat to be dissipated into the interior of the BC-10H for extra-cool operation.

OPERATING ECONOMY. Long tube life, low tube cost, and the highly efficient tank circuit combine to make economy of operation an important feature of the BC-10H.

ACCESSIBILITY. Designed for easy servicing, the transmitter front features 2 full-length doors, with operational controls located between the two. Meters which indicate transmitter operating parameters are located across the front of the cabinet, above the doors. All necessary tuning controls are adjustable in full view of these meters. Further access to the transmitter from the front may be gained by releasing the catches on various front access panels. In addition, 4 panels may be removed from the rear of the transmitter for 100% accessibility. The BC-10H is completely self-contained within one cabinet.

Front view, interior

INTERCHANGEABILITY. Added tube life may be achieved from the 3CX2500F3 triodes by interchanging the modulators and the power amplifiers, as the same tube type is used in both stages.

SOLID-STATE POWER SUPPLIES. Lifetime avalanche type silicon rectifiers in all power supplies provide a 2-to-1 voltage and a 5-to-1 current safety factor for normal operation and 150 times current ratings for surge currents. This high margin of safety assures trouble-free performance.

CONTROL CIRCUITRY. Careful attention has been given to the design of the control circuitry in the BC-10H. Complete AC and DC overload protection is standard equipment. A recycling feature, which will automatically turn the transmitter off when an overload occurs, is built in.

HARMONIC RADIATION. In addition to the full Pi-Tee output circuit, the BC-10H contains a second and a third harmonic trap to provide a bonus factor in meeting all harmonic attenuation requirements without relying on any other device in the system.

Fig. 2-21. (Cont.).

Rear view, interior

53

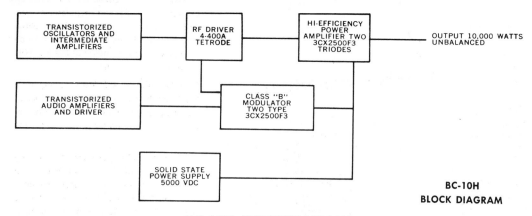

BC-10H
BLOCK DIAGRAM

BC-10H SPECIFICATIONS

POWER OUTPUT: (Rated) 10,000 watts. (Capable) 10,800 watts. Power reduction to 5,000, 2,500 or 1,000 watts available.

RF FREQUENCY RANGE: 535 kHz to 1620 kHz supplied to one frequency as ordered.

RF OUTPUT IMPEDANCE: Supplied for 50 ohms, or other as specified.

RF FREQUENCY STABILITY: ± 2 Hz.

CARRIER SHIFT: Less than 3% at 100% modulation.

RF HARMONICS: Meets or exceeds FCC specifications.

MODULATION CAPABILITY: Positive peaks 125%, negative peaks 100%.

AUDIO FREQUENCY RESPONSE: ± 1 dB, 50 to 10,000 Hz. ± 1½ dB, 30-12,000 Hz.

AUDIO FREQUENCY DISTORTION: 2.5% or less 50 Hz to 10,000 Hz at 95% modulation.

NOISE: (Unweighted) 60 dB or better below 100% modulation.

AUDIO INPUT: 600/150 ohms at + 10 dBm, ± 2 dB.

POWER INPUT: 208/230 volts, 3 phase, 50 or 60 Hz. 18.5 kW zero modulation. 21.0 kW average modulation. 27.5 kW 100% modulation.

AMBIENT TEMPERATURE RANGE: −20°C to +50°C.

ALTITUDE: To 7,500 ft standard (higher altitudes on special order).

SIZE: 78″ high, 72″ wide, 32″ deep (completely self-contained).

WEIGHT: 2,500 lbs. unpacked (approximate). 3,050 lbs. domestic packed (approximate). 3,250 lbs. export packed (approximate).

CUBAGE: 184 cubic feet packed.

FINISH: Beige-gray.

TUBES USED: (4) 3CX2500F3, (1) 4-400A. Total—5.

GENERAL INFORMATION: Monitors—10 RF volts output at 50/70 ohms for frequency and modulation monitors.

ORDERING INFORMATION

Model BC-10H transmitter with one set of tubes and two crystals 994-6522-005
100% set spare tubes for BC-10H transmitter 990-0539-001
Set of spare transistors for BC-10H [diodes not included] 990-0760-001
Kit for remote control of power output 994-6548-001

Fig. 2-21. (Cont.).

Corp. 10-kW transmitter. The following conclusions can be made from this information:

1. Tubes are still used for high-powered output stages.
2. Cooling features are a major factor in high-powered systems.
3. High-voltage dc power supplies are needed to obtain high-power outputs.
4. Operating economy is important in terms of easy serviceability, the use of highly derated components, and the use of energy-efficient circuits.

The inclusion of "dummy antennas" is a desirable feature in broadcast transmitters. A dummy antenna is a resistive load used in place of the antenna to prevent damage to the output circuits (which may occur under unloaded conditions) and prevents radiation of improper signals during tune-up and servicing if the regular antenna were used.

2-8 TRANSMITTER MEASUREMENTS

Trapezoid Patterns

A number of techniques are available to make operational checks on a transmitter's performance. A standard oscilloscope display of the transmitted AM signal will indicate any gross deficiencies. This technique is all the better if a dual-trace scope is available to allow the intelligence signal to be superimposed on the AM signal as illustrated in Fig. 2-13. An improvement in this method is known as the *trapezoidal pattern*. It is illustrated in Fig. 2-22. The AM signal is connected to the vertical input and the intelligence signal is applied to the horizontal input with the scope's internal sweep disconnected. The intelligence signal usually must be applied through an adjustable *RC* phase-shifting network, as shown, to ensure that it is exactly in phase with the modulation envelope of the AM waveform. Figure 2-22(b) shows the resulting scope display with improper phase relationships, and Fig. 2-22(c) shows the proper in-phase trapezoid pattern for a typical AM signal. It easily allows percentage modulation calculations by applying the B and A dimensions to the previously presented formula, Eq. (2-3).

$$\%m = \frac{B - A}{B + A} \times 100\% \qquad (2\text{-}3)$$

Figure 2-23 shows the trapezoidal pattern for a number of conditions and shows the corresponding standard AM waveform. In Fig. 2-23(a), the effect of 0% modulation (just the carrier) is indicated. The trapezoidal pattern is simply a vertical line, since there is no intelligence signal to provide horizontal deflection. At (b), normal modulation of about 50% has produced the trapezoidal pattern, while at (c), the condition of normal 100% modulation has extended the trapezoid into a triangular shape. The overmodulation condition at (d) causes the "sideways" triangle of 100% modulation to have a straight line off the right-hand side.

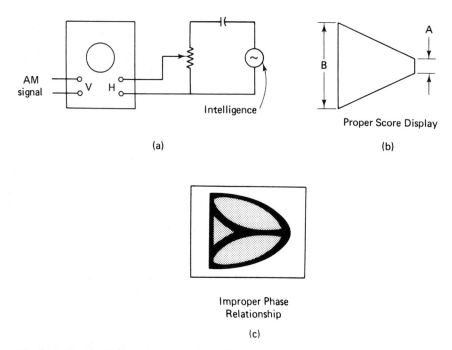

(a)

Proper Score Display

(b)

Improper Phase
Relationship

(c)

Fig. 2-22. Trapezoidal pattern connection scheme.

The "double-edged" pattern at (e) shows an effect of improper operation in the stage being modulated. The rise in output amplitude of the AM signal has been delayed and a slight distortion has been introduced in the modulation envelope. This condition is difficult to discern in the standard display, but the "double-edge" effect in the trapezoidal pattern is readily apparent. At (f), the effect of insufficient intelligence power is indicated. This is a rare case when the trapezoidal display does not pick up the problem, but the standard AM wave-envelope pattern obviously does. The wave envelope shows flattened positive and negative peaks.

Figure 2-24 shows two more trapezoidal displays indicative of some common problems. In both cases the trapezoid's sides are not straight (linear). The concave curvature of Fig. 2-24(a) indicates poor linearity in the modulation stage, which is often caused by improper neutralization or from stray coupling from a previous stage. The convex curvature at (b) is usually caused by improper bias or low carrier signal power (often termed low *excitation*).

Meter Measurement

It is possible to make some meaningful transmitter checks with a dc ammeter in the collector (or plate) of the modulated stage. If the operation is correct, this current should not change as the intelligence signal is varied between zero and the point where full modulation is attained. This is true since the increase in current

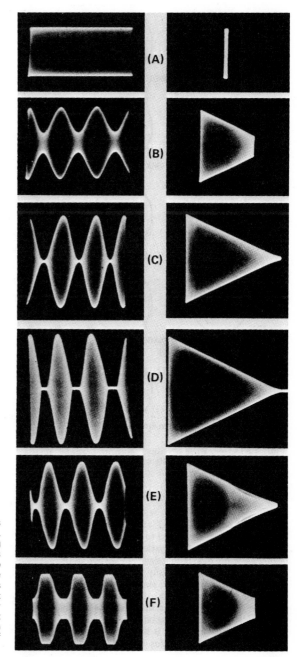

— Oscilloscope patterns showing various forms of modulation of an rf amplifier. At left, wave-envelope patterns; at right, corresponding trapezoidal patterns. The wave-envelope patterns were obtained with a linear oscilloscope sweep having a frequency one-third that of the sine-wave audio modulating frequency, so that three cycles of the modulation envelope may be seen. Shown at A is an unmodulated carrier, at B approximately 50-percent modulation, and at C, 100-percent modulation. The photos at D show modulation in excess of 100 percent. E and F show the results of improper operation or circuit design. See text.

Fig. 2-23. Trapezoidal and standard AM displays for varying conditions. (Courtesy of The ARRL Radio Amateur's Handbook.)

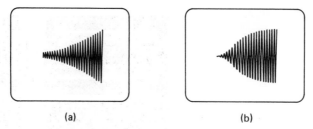

(a)   (b)

Fig. 2-24. Trapezoid patterns indicating problems.

during the crest of the modulated wave should be exactly offset by the drop during the trough. A distorted AM signal will usually cause a change in dc current flow. In the case of overmodulation, the current will increase further during the crest but cannot decrease below zero at the trough, and a net increase in dc current will occur. It is also common for this current to decrease as modulation is applied. This malfunction is termed *downward modulation* and is usually the result of insufficient excitation. The current increase during the modulation envelope crest is minimized, but the decrease during the trough is nearly normal.

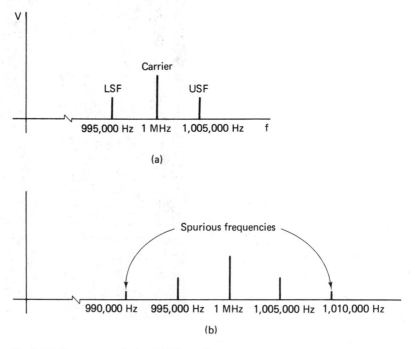

Fig. 2-25. Spectrum analysis of AM waveforms.

Spectrum Analyzers

The use of spectrum analyzers in recent years has shown a dramatic increase in all fields of electronics, but especially in the communications industry. A *spectrum analyzer* visually displays (on a CRT) the amplitude of the components of a wave as a function of frequency. This can be contrasted with an oscilloscope display which shows the amplitude of the total wave (all components) versus time. Thus, an oscilloscope shows us the "time" domain while the spectrum analyzer shows the "frequency" domain. In Fig. 2-25(a) the frequency domain for a 1-MHz carrier modulated by a 5-kHz intelligence signal is shown. Proper operation is indicated since only the carrier and upper and lower side frequencies are present. During malfunctions, and to a lesser extent even under normal conditions, transmitters will often generate *spurious frequencies* as shown at (b), where components other than just the three desired are present. These spurious undesired components are usually termed simply *spurs*, and their amplitude is severely controlled by FCC regulation to minimize interference on adjacent channels. The coupling stage between the transmitter and its antenna is designed to attenuate the spurs, but the transmitter's output stage must also be carefully designed to keep the spurs to a minimum level. The use of spectrum analyzers is obviously a very handy tool for use in evaluating a transmitter's performance.

3

AMPLITUDE MODULATION:

RECEPTION

3-1 RECEIVER CHARACTERISTICS

If you were to logically envision a block diagram for a radio receiver, you would probably go through the following thought process:

1. The signal from the antenna is usually very small—therefore, amplification is necessary at the frequency of the carrier. This stage should have low-noise characteristics and should be "tuned" to accept only the desired carrier and side-band frequencies to avoid interference from other stations and to minimize the received noise since external noise is proportional to bandwidth.
2. After sufficient amplification, a circuit to extract the intelligence from the radio frequency is necessary.
3. Following the detection of the intelligence, further amplification is necessary to give it sufficient power to drive a loudspeaker.

This logical train of thought leads to the block diagram shown in Fig. 3-1. It consists simply of an RF amplifier, detector, and audio amplifier. The first radio

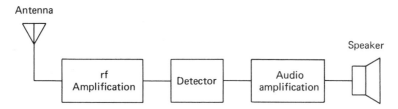

Fig. 3-1. Simple radio receiver block diagram.

receivers for broadcast AM took this form and are called *tuned radio frequency* or, more simply, *TRF receivers*. These receivers generally had three stages of RF amplification, with each stage preceded by a separate variable-tuned circuit. You can imagine the frustration experienced by the user when tuning to a new station. The three tuned circuits were all adjusted by separate variable capacitor controls. To receive a station required proper adjustment of all three, and a good deal of time and practice was necessary.

Sensitivity and Selectivity

Two major characteristics of any receiver are its sensitivity and selectivity. A receiver's *sensitivity* may be defined as its ability to drive its output transducer to an acceptable level. A more technical definition is the minimum input signal (usually expressed as a voltage) required to produce a specified output signal. The range of sensitivities for communication receivers varies from the millivolt region for low-cost AM receivers down to the nanovolt region for ultrasophisticated units for more exacting applications. In essence, a receiver's sensitivity is determined by the amount of gain provided and, more important, its noise characteristics. It is not difficult to insert more gain in a radio, but getting noise figures below a certain level becomes a very exacting science.

Selectivity may be defined as the extent to which a receiver is capable of differentiating between the desired signal and disturbances at other frequencies (unwanted radio signals and noise). A receiver can also be overly selective. For instance, on commercial broadcast AM, we have seen that the transmitted signal can handle intelligence signals up to a maximum of 5 kHz, which subsequently generates upper and lower side bands extending 5 kHz above and below the carrier frequency. Thus, the total signal has a 10-kHz bandwidth. Optimum receiver selectivity is thus 10 kHz, but if a 5-kHz bandwidth were "selected," the upper and lower side bands would only extend 2.5 kHz above and below the carrier. The radio's output would suffer from a lack of the full possible "fidelity," since the output would include intelligence up to a maximum of 2.5 kHz. On the other hand, an excessive selectivity of 20 kHz results in the reception of unwanted adjacent radio signals and the additional external noise that is directly proportional to the bandwidth selected. A glance at Fig. 3-2 will help to illustrate these situations of over- and underselectivity. Unfortunately, TRF receivers did suffer from problems

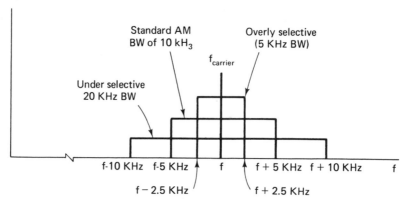

Fig. 3-2. Illustrations of over and under selectivity.

of poor selectivity. It was this problem that led to their replacement by the super-heterodyne receiver.

LC-Circuit Considerations

It is necessary to consider some characterisitcs of tuned *LC* circuits to get a good grasp of this selectivity problem. Figure 3-3 shows a tuned circuit and a typical output/input-versus-frequency characteristic. Notice that the output is maximum at the tank circuit's resonant frequency f_o, which is predicted with sufficient accuracy by

$$f_o = \frac{1}{2\pi\sqrt{LC}} \tag{3-1}$$

At frequencies above and below f_o, the circuit's output falls off since its impedance is maximum at f_o. Its high-frequency cutoff (f_{hc}) and low-frequency cutoff (f_{lc}) are

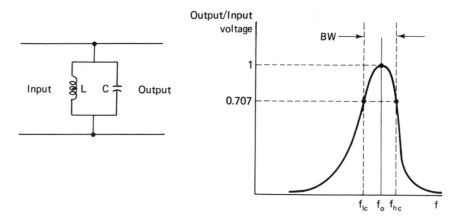

Fig. 3-3. LC tuned circuit and its output/input characteristics.

defined as the points where the output voltage is down to 0.707 of the maximum output at f_o. This corresponds to the point where the power out is one-half maximum since power is proportional to voltage squared $[(0.707)^2 = 0.5]$. Expressed in decibel form,

$$V_{\text{out}} = 0.707 V_{\text{IN}}$$

$$\frac{V_{\text{out}}}{V_{\text{in}}} = 0.707$$

$$\frac{V_{\text{out}}}{V_{\text{in}}} \text{ dB} = 20 \log \frac{V_{\text{out}}}{V_{\text{in}}} \tag{3-2}$$

or

$$\text{dB} = -20 \log \frac{V_{\text{in}}}{V_{\text{out}}} \tag{3-3}$$

$$= -20 \log \frac{1}{0.707}$$

$$= -20 \log 1.414$$

$$= -3 \, \text{dB}$$

Thus, the high- and low-frequency cutoffs are often referred to as the *3-dB down points*. The same result is obtained when using the power ratio instead of the voltage ratio.

$$P_{\text{out}} = 0.5 P_{\text{in}}$$

$$\frac{P_{\text{out}}}{P_{\text{in}}} = 0.5$$

$$\frac{P_{\text{out}}}{P_{\text{in}}} \text{ dB} = 10 \log \frac{P_{\text{out}}}{P_{\text{in}}} \tag{3-4}$$

or

$$\text{dB} = -10 \log \frac{P_{\text{in}}}{P_{\text{out}}} \tag{3-5}$$

$$= -10 \log \frac{1}{0.5}$$

$$= -10 \log 2$$

$$= -3 \, \text{dB}$$

The minus sign in decibel ratios indicates a loss or attenuation, while a positive decibel ratio shows signal gain or amplification. Whenever the ratio $P_{\text{out}}/P_{\text{in}}$ or $V_{\text{out}}/V_{\text{in}}$ is less than 1, take the log of the reciprocal ($P_{\text{in}}/P_{\text{out}}$ or $V_{\text{in}}/V_{\text{out}}$) and put a negative sign in front of the result to indicate an attenuation or loss.

The *bandwidth* of a tuned circuit is the range of frequency included between the high- and low-frequency cutoffs:

$$\text{bandwidth (BW)} = f_{\text{hc}} - f_{\text{lc}} \tag{3-6}$$

Another important parameter for tuned circuits is the *quality factor*, or *Q*. The *Q* provides a measure of the tuned circuit's selectivity. The higher the *Q*, the narrower will be the BW. The relationship between *Q*, BW, and f_o, is

$$Q = \frac{f_o}{\text{BW}} \qquad (3\text{-}7)$$

The *Q* of a tuned circuit is primarily determined by how purely inductive and capacitive are its elements. Since pure high-quality capacitors are easily fabricated, it is usually the inductor's winding resistance that is the dominant factor in *Q* determination.

TRF Selectivity

Now, getting back to the selectivity problem of TRF receivers, this information on tuned circuits will be applied. Consider a standard AM broadcast band receiver that spans the frequency range from 550 kHz to 1550 kHz. If the approximate center of this frequency range of 1000 kHz is considered, we find that the desired 10-kHz BW requires a *Q* of 100.

$$Q = \frac{f_o}{\text{BW}} \qquad (3\text{-}7)$$

$$= \frac{1000 \text{ kHz}}{10 \text{ kHz}}$$

$$= 100$$

Now, since the *Q* of a tuned circuit remains fairly constant as its capacitance is varied, a change to 1550 kHz will change the BW to 15.5 kHz.

$$Q = \frac{f_o}{\text{BW}} \qquad (3\text{-}7)$$

Therefore,

$$\text{BW} = \frac{f_o}{Q} \qquad (3\text{-}8)$$

$$= \frac{1550 \text{ kHz}}{100}$$

$$= 15.5 \text{ kHz}$$

The receiver's BW is now too large, and it will suffer from adjacent station interference and increased noise. On the other hand, the opposite problem is encountered at the lower end of the frequency range. At 550 kHz, the BW is 5.5 kHz.

$$\text{BW} = \frac{f_o}{Q} \qquad (3\text{-}8)$$

$$= \frac{550 \text{ kHz}}{100}$$

$$= 5.5 \text{ kHz}$$

The fidelity of reception is now impaired. The maximum intelligence frequency possible is 5.5 kHz/2 or 2.75 kHz instead of the full 5 kHz transmitted. It is this selectivity problem that led to the general use of the superheterodyne receiver in place of TRF designs.

EXAMPLE 3-1

A TRF receiver is to be designed with a single tuned circuit using a 10-μH inductor.

    A. Calculate the capacitance range of the variable capacitor required to tune the 550- to 1550-kHz range.
    B. The ideal 10-kHz BW is to occur at 1100 kHz. Determine the required Q.
    C. Calculate the BW of this receiver at 550 kHz and 1550 kHz.

Solution:

    A. At 550 kHz, calculate C as

$$f_o = \frac{1}{2\pi\sqrt{LC}} \qquad (3\text{-}1)$$

$$550\,\text{kHz} = \frac{1}{2\pi\sqrt{10\,\mu\text{H} \times C}}$$

$$C = 8.37\,\text{nF}$$

At 1550 kHz,

$$1550\,\text{kHz} = \frac{1}{2\pi\sqrt{10\,\mu\text{H} \times C}}$$

$$C = 1.06\,\text{nF}$$

Therefore, the required range of capacitance is from

$$1.06\,\text{nF} \quad \text{to} \quad 8.37\,\text{nF}$$

    B. $Q = \dfrac{f_o}{\text{BW}} \qquad (3\text{-}7)$

$$= \frac{1100\,\text{kHz}}{10\,\text{kHz}}$$

$$= 110$$

    C. At 1550 kHz,

$$\text{BW} = \frac{f_o}{Q} \qquad (3\text{-}8)$$

$$= \frac{1550\,\text{kHz}}{110}$$

$$= 14.1\,\text{kHz}$$

At 550 kHz,

$$\text{BW} = \frac{550\,\text{kHz}}{110}$$

$$= 5\,\text{kHz}$$

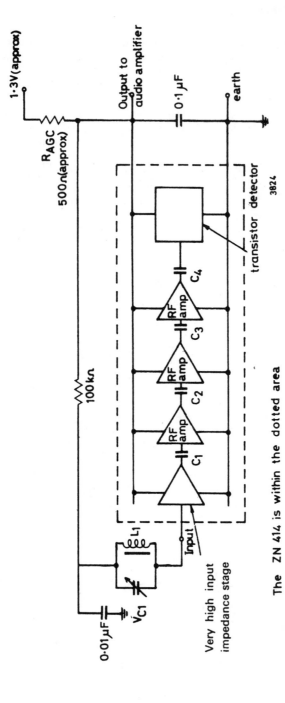

The ZN 414 is within the dotted area

Fig. 3-4. The Ferranti Limited ZN414 TRF receiver LIC. (Courtesy of the Ferranti Electonic Components Division.)

LIC TRF Receiver

In spite of the selectivity problem, the use of TRF designs does find application in receivers designed for single-frequency or narrow-band applications. It offers excellent quality and extreme simplicity and subsequent low cost. Ferranti Ltd. offers a TRF receiver in a single three-leaded LIC incorporated within a TO-18 transistor case. Figure 3-4 shows its block diagram along with the single required external tuned circuit and the biasing setup. The ZN414 includes a high input impedance stage along with three capacitively coupled RF amplifiers and a transistor detector to extract the intelligence from the carrier and side bands. The output shown could drive a sensitive earphone or subsequent audio amplification can be utilized to drive a speaker. The R_{AGC} resistor provides the automatic gain control function to provide a relatively constant output signal under varying input signal conditions. More detail on this subject will be provided later in this chapter. Figure 3-5 provides some of the important specifications of the ZN414. Notice that it can be used for carrier frequencies from about 150 kHz to 3 MHz, has an approximate sensitivity of 50 μV, has a typical power gain of 72 dB and can compensate for roughly 20 dB of input signal variation with its automatic gain control feature.

EXAMPLE 3-2

Provide a block diagram of a TRF receiver using the ZN414 for an AM station at 1210 kHz. A 15.9-nF capacitor is to be in the input tuning circuit and the output should exceed 300 mV.

Solution:

Since a fixed 1210-kHz carrier is the design goal, the first step is to calculate the required inductance:

$$f_o = \frac{1}{2\pi\sqrt{LC}} \qquad (3\text{-}1)$$

$$1210 \text{ kHz} = \frac{1}{2\pi\sqrt{L \times 15.9 \text{ pF.}}}$$

$$L = 1.09 \text{ mH}$$

Next, the Q of the tank circuit should be determined to provide the desired 10-kHz BW for standard broadcast AM.

$$Q = \frac{f_o}{\text{BW}} \qquad (3\text{-}7)$$

$$= \frac{1210 \text{ kHz}}{10 \text{ kHz}}$$

$$= 121$$

Since it is required to supply 300 mV of output and the ZN414 is

SUMMARY OF PARAMETERS

Supply voltage range	1·2-1·6 volts (1·3 volts recommended)
Storage temperature range	–65°C to +125°C
Operating temperature range	0°C to +70°C
Supply current	... 0·3 mA typical (0·5mA under strong signal conditions)
Frequency range	... 150kHz-3MHz useful range
Input resistance	... 4MΩ typical
Threshold sensitivity	... 50μV with 1·3 volt supplies (dependent on 'Q' of coil)
Audio distortion	... ⩽ 2% T.H.D. under correct operating conditions
Selectivity	... 4kHz bandwidth can be achieved
Power gain	... 72dB typical
AGC range	... 20dB typical (dependent on R_{AGC})
Output	⩾ 30mV r.m.s. under correct operating conditions

ZN414 CHARACTERISTICS All measurements performed with 30% modulation, f_M = 400Hz

Gain and AGC characteristics

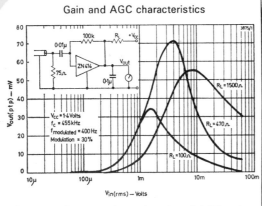

See operating notes for explanation of AGC action.

Frequency response of the ZN414

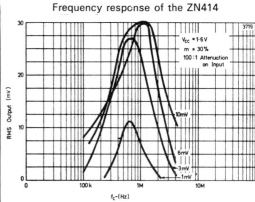

Note that this graph represents the chip response, and not the receiver bandwidth.

Gain variation with supply volts

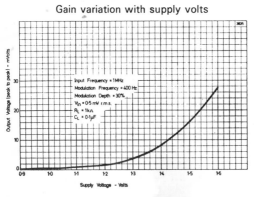

D.C. level at output

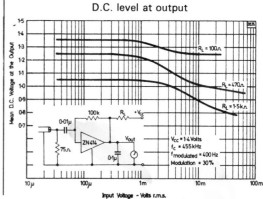

Fig. 3-5. ZN414 specifications. (Courtesy of the Ferranti Electronic Components Division.)

68

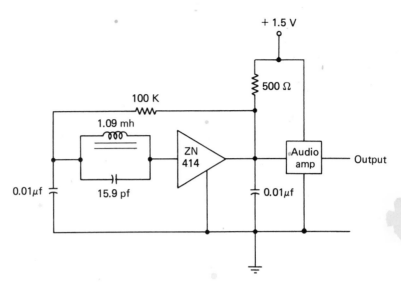

Fig. 3-6. Solution for Example 3-2.

guaranteed to provide only 30 mV, some additional amplification is required. Thus, Fig. 3-6 provides the requested block diagram.

3-2 AM DETECTION

The process of detecting the intelligence out of the carrier and side bands (the AM signal) has thus far been mentioned as a function but not explained in detail. In fact, the detection process can be very simply accomplished. Recall our discussions about generating AM. We said if two different frequencies were passed through a nonlinear device, that sum and difference components would be generated. The carrier and side bands of the AM signal are separated in frequency by an amount equal to the intelligence frequency and are, thus, sum and difference frequencies. If the complete AM signal is passed through a nonlinear device, difference frequencies between the carrier and side bands will be generated and these frequencies are, in fact, the intelligence. It follows, then, that passing the AM signal through a nonlinear device will enable the process of detection, just as passing the carrier and intelligence through a nonlinear device enables AM generation.

Detection of amplitude-modulated signals requires a nonlinear electrical network. An ideal nonlinear curve for this is one that affects the positive half-cycles of the modulated wave differently from the negative half-cycles, and that so distorts an applied voltage wave of zero average value that the average resultant current varies as the intelligence signal amplitude. The curve shown in Fig. 3-7 is called an *ideal curve* because it is linear on each side of the operating point *P* and does not introduce harmonic frequencies.

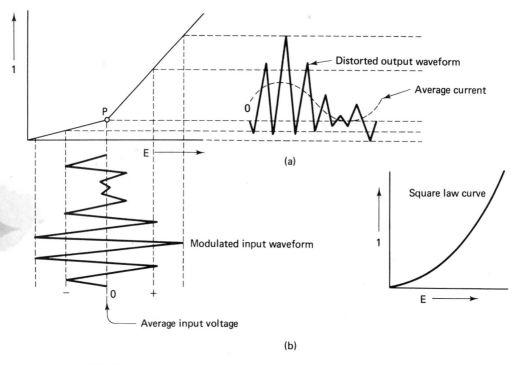

Fig. 3-7. Nonlinear device used as detector.

When the input to an ideal nonlinear curve is a carrier and its side bands, the output contains the following frequencies:

1. The carrier frequency⎫
2. The upper side band ⎬original components.
3. The lower side band ⎭
4. A dc component.
5. A frequency equal to the carrier minus the lower side band (or a frequency equal to the upper side band minus the carrier), which is the original signal frequency.

The detector reproduces the signal frequency by producing a distortion of a desirable kind in its output. When the output of the detector is impressed upon a low-pass filter, which suppresses the radio frequencies, only the original or signal frequency is left. This is shown as the dotted average current curve in Fig. 3-7(a).

In some practical detector circuits, the nearest approach to the ideal curve is the square-law curve shown in Fig. 3-8(b). The output of a device using this curve contains, in addition to all the frequencies that were listed, the harmonics of each of the input frequencies. The harmonics occur because voltage inputs that have a

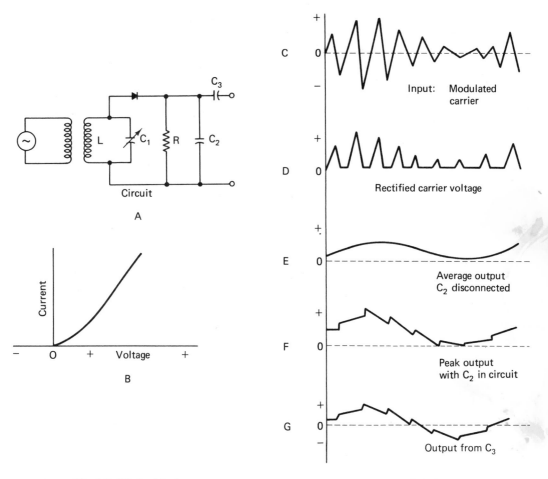

Fig. 3-8. Diode detector.

large amplitude are distorted differently from voltage inputs that have a low amplitude. The harmonics of radio frequencies can be filtered out, but the harmonics of the sum and difference frequencies, even though they produce an undesirable distortion, have to be tolerated, since they are in the audio-frequency range.

Diode Detector

One of the simplest and most effective types of detectors, and one with nearly an ideal nonlinear resistance characteristic, is the diode detector circuit shown in Fig. 3-8(a). Notice the *I–V* curve in Fig. 3-8(b). This is the type of curve on which the diode detector at *A* operates. The curved part of the curve is the region of low current and indicates that for small signals the output of the detector will follow the square law. For input signals with large amplitudes, however, the output is essentially linear and, therefore, harmonic outputs are limited.

The modulated carrier is introduced into the tuned circuit made up of LC_1 in Fig. 3-8(a). The waveshape of the input to the diode is shown in Fig. 3-8(c). As a diode conducts only during half-cycles, this circuit removes all the negative half-cycles and gives the result shown in Fig. 3-8(d). The average output is shown at E. Although the average input voltage is zero, the average output voltage across R always varies above zero.

The low-pass filter, made up of capacitor C_2 and resistor R, removes the RF (carrier frequency), which, so far as the rest of the receiver is concerned, serves no useful purpose. Capacitor C_2 charges rapidly to the peak voltage through the small resistance of the conducting diode, but discharges slowly through the high resistance of R. The sizes of R and C_2 normally form a rather short time constant at the intelligence (audio) frequency and a very long time constant at the radio frequencies. The resultant output with C_2 in the circuit is a varying voltage that follows the peak variation of the modulated carrier [see Fig. 3-8(f)]. The dc component produced by the detector circuit is still in the waveshape but may be removed by capacitor C_3, producing the ac voltage waveshape in Fig. 3-8(g). In communications receivers the dc component is often used for providing automatic volume (gain) control.

Advantages of diode detectors are as follows:

1. Ability to handle relatively high power signals. There is no practical limit to the amplitude of the input signal.
2. Distortion levels acceptable for most AM applications. Distortion decreases as the amplitude increases.
3. High efficiency. When properly designed, 90% efficiency is obtainable.
4. Diode detectors develop a readily usable dc voltage for the automatic gain control circuits.

Disadvantages of the diode detectors are:

1. Power is absorbed from the tuned circuit by the diode circuit. This reduces the Q and selectivity of the tuned input circuit.
2. No amplification occurs in a diode detector circuit.

Diagonal Clipping

Careful selection of component parts is necessary for obtaining optimum efficiency in diode detector circuits. One very important fact to consider is the value of the time constant RC_2, particularly in the case of pulse modulation. When a carrier modulated by a square pulse [Fig. 3-9(b)] is applied to an ideal diode detector, the waveshape shown in Fig. 3-9(c) is produced. Notice that for clarity, the amplitude of the wave at C is exaggerated in comparison to the high-frequency carrier shown at B.

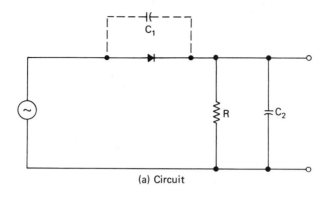

(a) Circuit

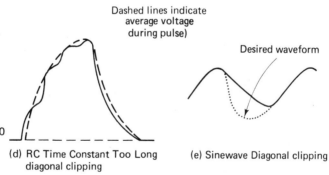

(b) Input to Detector

(c) Ideal RC Time Constant

Dashed lines indicate
average voltage
during pulse)

Desired waveform

(d) RC Time Constant Too Long
diagonal clipping

(e) Sinewave Diagonal clipping

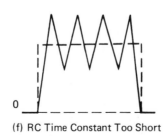

(f) RC Time Constant Too Short

Fig. 3-9. Diode detector component considerations.

73

If the time constant of RC_2 is too long, several cycles are required to charge C_2, and the leading edge of the output pulse is sloped as shown in Fig. 3-9(d). After the pulse passes by, the capacitor charges slowly and the trailing edge is exponential rather than square as desired. This phenomenon is often referred to as "diagonal clipping." The diagonal clipping effect from a sine-wave intelligence signal is shown at E. Notice that the detected sine wave at E is distorted. The excessive RC time constant did not allow the capacitor voltage to follow the full changes of the sine wave. If the time constant is too short, both the leading and trailing edges can be easily reproduced. However, the capacitor may discharge considerably between cycles, and this reduces the average amplitude of the pulse, leaving a sizable component of the carrier frequency in the output, as shown in Fig. 3-9(f).

For these reasons the selection of the time constant is a compromise. To achieve this compromise, the load resistor R must be large, as the total input voltage is divided across R and the internal resistance of the diode when it is conducting. A large value of load resistance ensures that the greater part of this voltage will be in the output, where it is desired. On the other hand, the load resistance must not be so high that capacitor C_2 becomes small enough to approximate the size of C_i [Fig. 3-9(a)], the internal junction capacitance of the diode. When this occurs, capacitor C_2 will try to discharge through C_i during the nonconducting periods, which would reduce the amplitude of the detector output.

Synchronous Detection

Diode detectors are used in the vast majority of AM detection schemes. Since "high fidelity" is usually not an important aspect in AM, the distortion levels of several percentage points or more from a diode detector can easily be tolerated. In applications demanding greater performance, the use of a synchronous detector offers the following advantages:

1. Low distortion—well under 1%.
2. Greater ability to follow fast-modulation waveforms, as in pulse-modulation or high-fidelity applications.
3. The ability to provide gain instead of the attenuation of diode detectors.

Since synchronous detectors require rather complex circuitry, the use of a LIC is most appropriate. A LIC offered by RCA (CA3067) or National Semiconductor (LM3067) was designed specifically for use as a TV chroma (color) demodulator. It is easily adapted for use as a synchronous AM detector. Figure 3-10 shows the circuit connection. The tint amplifier provides an initial stage of gain with the double-balanced demodulators providing the synchronous detection. With a 35-mV AM signal input, a 450-mV audio output is obtained with less than 0.7% distortion at 80% modulation. The circuit can be used with carrier frequencies as

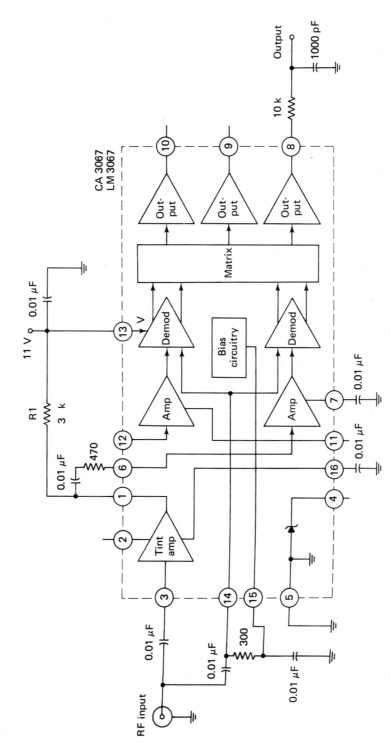

Fig. 3-10. Synchronous AM detection.

low as 10 kHz and up to 10 MHz with higher-frequency operation made possible by substituting a tuned circuit for R_1, which provides proper adjustment of the carrier phase.

3-3 SUPERHETERODYNE RECEIVERS

Tuned Radio Frequency

The basic problem of variable selectivity in TRF systems led to the development and general usage of the superheterodyne receivers in the early 1930s. This basic receiver configuration is still dominant after all these years, which provides an indication of its utility. A block diagram for a superheterodyne receiver is provided in Fig. 3-11. The first stage shown, in dotted lines, is a standard RF amplifier which may or may not be required, depending upon factors to be discussed later. The next stage is the mixer, which accepts two inputs, the output of the RF amplifier (or antenna input when an RF amplifier is omitted) and a steady sine wave from the local oscillator (LO). The mixer is yet another nonlinear device utilized in AM. Its function is to mix the AM signal with a sine wave to generate a new set of sum and difference frequencies. Its output, as will be shown, is an AM signal with a constant carrier frequency regardless of the frequency of the station tuned to. The next stage is the intermediate-frequency (IF) amplifier, which provides the bulk of radio-frequency signal amplification at a *fixed* frequency. This allows for a constant BW

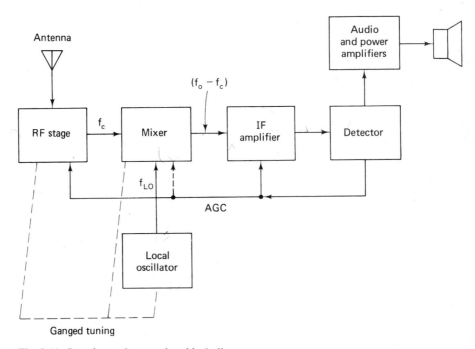

Fig. 3-11. Superheterodyne receiver block diagram.

regardless of the station tuned and is the key to the superior selectivity of the superheterodyne receiver. Following the IF amplifiers is the detector, which extracts the intelligence from the radio signal. It is subsequently amplified by the audio amplifiers into the speaker. A dc level proportional to the received signal's strength is extracted from the detector stage and fed back to the IF amplifiers and sometimes even the mixer and/or the RF amplifier. This is the automatic gain control (AGC) level, which allows relatively constant receiver output for widely variable radio signals.

Frequency Conversion

It has been stated that the mixer performs a frequency conversion process. Consider the situation shown in Fig. 3-12. The AM signal into the mixer is a 1000-kHz carrier that has been modulated by a 1-kHz sine wave, thus producing side frequencies at 999 kHz and 1001 kHz. The LO input is a 1455-kHz sine wave. The mixer, being a nonlinear device, will generate the following components:

1. Frequencies at all of the original inputs: 999 kHz, 1000 kHz, 1001 kHz, and 1455 kHz.
2. Sum and difference components of all the original inputs: 1455 kHz $\pm$ (999 kHz, 1000 kHz, and 1001 kHz). This means outputs at 2454 kHz, 2455 kHz, 2456 kHz, 454 kHz, 455 kHz, and 456 kHz.
3. Harmonics of all the frequencies listed in components 1 and 2 and a dc component.

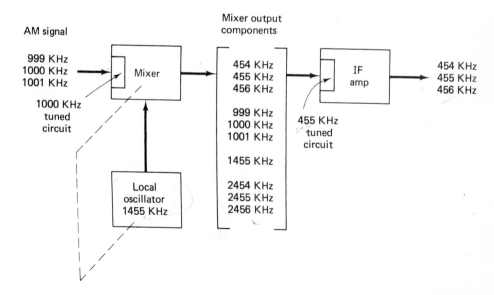

Fig. 3-12. The frequency conversion process.

The first IF amplifier has a tuned circuit that accepts only components near 455 kHz, in this case 454 kHz, 455 kHz, and 456 kHz. Since the mixer maintains the same amplitude proportion that existed with the original AM signal input at 999 kHz, 1000 kHz, and 1001 kHz, the signal now passing through the IF amplifiers is a replica of the original AM signal. The only difference is that now its carrier frequency is 455 kHz, but its envelope is identical to that of the original AM signal. A frequency conversion has occurred that has translated the carrier from 1000 kHz to 455 kHz—a frequency intermediate to the original carrier and intelligence frequencies—which led to the terminology "intermediate-frequency amplifier," usually called the IF amplifier. Since the mixer and detector both have the nonlinear characteristic, the mixer is often referred to as the *first detector*.

Tuned-Circuit Adjustment

Now consider the effect of changing the tuned circuit at the front end of the mixer to accept a station at 1600 kHz. This means a reduction in either its inductance or capacitance (usually the latter) to change its center frequency from 1000 kHz to 1600 kHz. If the capacitance in the local oscillator's tuned circuit were simultaneously reduced so that its frequency of oscillation went up by 600 kHz, the situation shown in Fig. 3-13 would now exist. The mixer's output still contains a component at 455 kHz (among others) as in the previous case, when we were tuned to a 1000-kHz station. Of course, the other frequency components at the output of the mixer are not accepted by the selective circuits in the IF amplifiers.

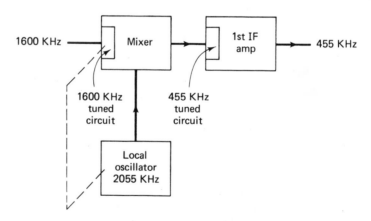

Fig. 3-13. Frequency conversion.

Thus, the key to superheterodyne operation is to make the LO frequency "track" with the circuit or circuits that are tuning the incoming radio signal such that their difference is a constant frequency (the IF). For a 455-kHz IF frequency, the most common case for broadcast AM receivers, this means the LO should always be at a frequency 455 kHz above the incoming carrier frequency. The receiver's "front end" tuned cricuits are usually made to track together by mechanically linking (ganging) the capacitors in these circuits on a common variable rotor assembly, as shown in Fig. 3-14. Note that this ganged capacitor has three separate capacitor elements.

Trimmers

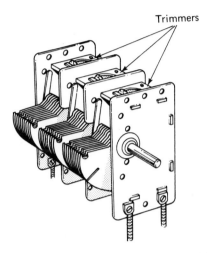

Fig. 3-14. Variable ganged capacitor.

3-4 SUPERHETERODYNE TUNING

Tracking

It is not possible to make a receiver track perfectly over an entire wide range of frequencies. The perfect situation occurs when the RF amplifier and mixer tuned circuits are exactly together and the LO is above these two by an amount exactly equal to the IF frequency. To obtain a practical degree of tracking, the following steps are employed:

1. A small variable capacitance in parallel with each section of the ganged capacitor, called the *trimmer*, is adjusted for proper operation at the highest frequency. The trimmer capacitors are shown in Fig. 3-14. The highest frequency requires the main capacitor to be at its minimum value (i.e., the plates all the way open). The trimmers are then adjusted to balance out the remaining stray capacitances to provide perfect tracking at the highest frequency.

2. At the lowest frequency, when the ganged capacitors are fully meshed and thus maximum, a small variable capacitor known as the *padder* capacitor is put in series with the tank inductor. The "padders" are adjusted to provide tracking at the low frequency in the band.

3. The final adjustment is made at midfrequency by slight adjustment of the inductance in each tank.

The curve in Fig. 3-15(a) shows that performing the steps above, and then rechecking them once again to allow for interaction effects, provides perfect

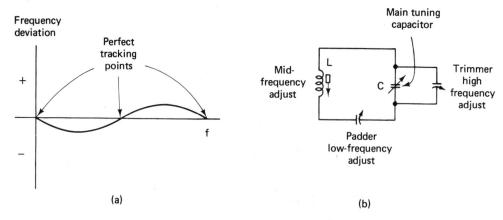

Fig. 3-15. Tracking considerations.

tracking at three points. The minor imperfections between these points are generally of an acceptable nature. Figure 3-15(b) shows the circuit for each tank circuit and summarizes the adjustment procedure.

Electronic Tuning

The bulk and cost of ganged capacitors has led to their gradual replacement by a technique loosely called "electronic tuning". The process relies on the capacitance offered by a reverse-biased diode. Since this capacitance varies with the amount of reverse bias, a potentiometer can be used to provide the variable

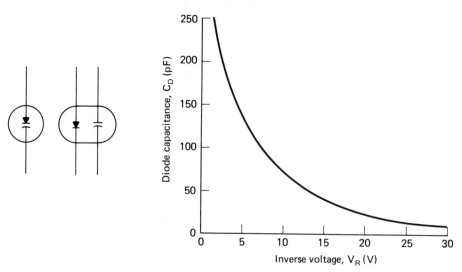

Fig. 3-16. Varactor diode symbols and C/V characteristic.

capacitance required for tuning. Diodes that have been specially fabricated to enhance this variable capacitance versus reverse bias characteristic are referred to as *varactor diodes, varicap diodes,* or *VVC diodes.* Figure 3-16 shows the two generally used symbols for these diodes and a typical capacitance-versus-reverse bias characteristic.

Figure 3-17 shows the front end of a broadcast band receiver. It does not incorporate an RF amplifier, and Q_1 performs the dual function of mixer and the local oscillator. The varactor diode D_1 provides the variable capacitance necessary to tune the radio signal from the antenna while D_2 allows for the variable LO frequency. The -1- to -12-V supply comes from the tuning potentiometer and provides the necessary variable reverse voltage for both varactor diodes. The matched diode characteristics required for good tracking often lead to the use of varactor diodes fabricated on a common silicon chip and provided in a single package.

Besides the previously mentioned advantage of bulk and cost savings, electronic tuning is not as prone to dust and humidity problems. It is also more easily adapted to remote-control and pushbutton tuning. The costly mechanical pushbutton systems can be replaced by simple electrical switches connected to a resistive voltage divider with adjustable taps.

3-5 SUPERHETERODYNE ANALYSIS

Image Frequency

The superheterodyne receiver has been shown to have that one great advantage over the TRF—constant selectivity over a wide range of received frequencies. This was shown to be true since the bulk of the amplification in a superheterodyne receiver occurs in the IF amplifiers at a fixed frequency, and this allows for relatively simple and yet highly effective frequency selective circuits. A disadvantage does exist, however, other than the obvious fact that a superheterodyne receiver is somewhat more complex. The frequency conversion process performed by the mixer–oscillator combination sometimes will allow a station other than the desired one to be fed into the IF and subsequently amplified. Consider a receiver tuned to receive a 20-MHz station that uses a 1-MHz IF. The LO would, in this case, be at 21 MHz to generate a 1-MHz frequency component at the mixer output. This situation is illustrated in Fig. 3-18. If an undesired station at 22 MHz were also on the air, it is possible for it to also get into the mixer. Even though the tuned circuit at the mixer's front end is "selecting" a center frequency of 20 MHz, a look at its response curve in Fig. 3-18 shows that it will not fully attenuate the undesired station at 22 MHz. As soon as the 22-MHz signal is fed into the mixer, we have a problem. It mixes with the 21-MHz LO signal and one of the components produced is 22 MHz $-$ 21 MHz $=$ 1 MHz—the IF frequency! Thus, we now have a desired 20-MHz station and an undesired 22 MHz station, which both look

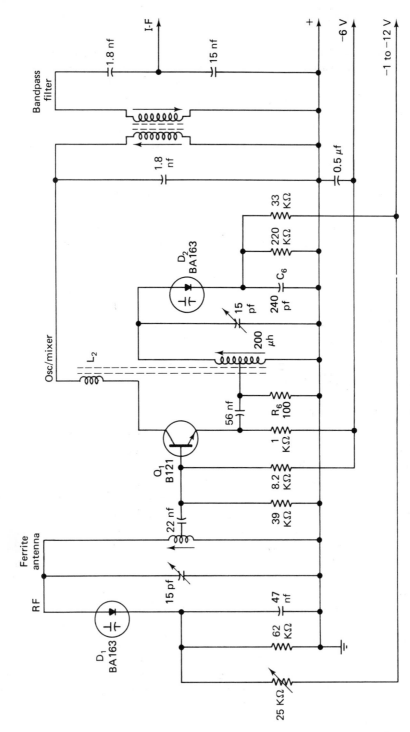

Fig. 3-17. Broadcast band AM receiver, front end, with electronic tuning.

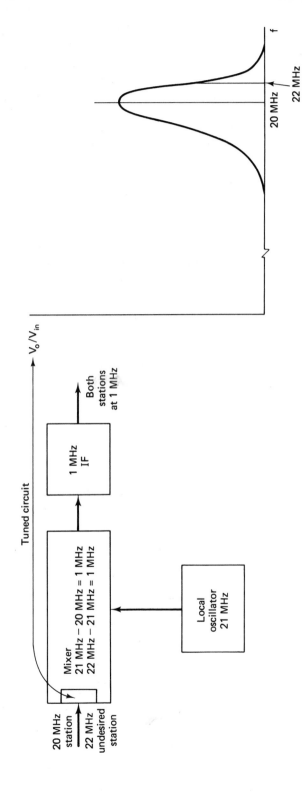

Fig. 3-18. Image frequency illustration.

correct to the IF. Depending on the strength of the undesired station, it can either interfere or even completely override the desired station.

In Ex. 3-2, the undesired 22-MHz station is called the *image frequency*. The designing of superheterodyne receivers with a high degree of image frequency rejection is obviously an important consideration.

EXAMPLE 3-3

Determine the image frequency for a standard broadcast band receiver using a 455-kHz IF and tuned to a station at 620 kHz.

Solution:

The first step is to determine the frequency of the LO. The LO frequency minus the desired station's frequency of 620 kHz should equal the IF of 455 kHz. Hence,

$$\text{LO} - 620 \text{ kHz} = 455 \text{ kHz}$$
$$\text{LO} = 620 \text{ kHz} + 455 \text{ kHz}$$
$$= 1075 \text{ kHz}$$

Now determine what other frequency, when mixed with 1075 kHz, yields an output component at 455 kHz.

$$X - 1075 \text{ kHz} = 455 \text{ kHz}$$
$$X = 1075 \text{ kHz} + 455 \text{ kHz}$$
$$= 1530 \text{ kHz}$$

Thus, 1530 kHz is the image frequency in this situation.

Example 3-3 shows that image frequency rejection on the standard broadcast band is not a major problem. A glance at Fig. 3-19 serves to illustrate this point.

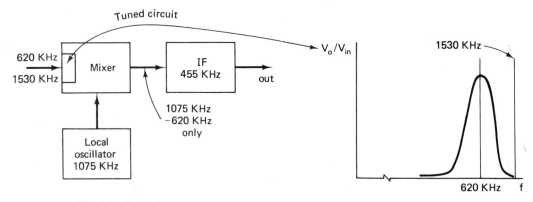

Fig. **3-19.** Image frequency not a problem.

The tuned circuit at the mixer's input comes fairly close to fully attenuating the image frequency, in this case, since 1530 kHz is so far away from the tuned circuit's center frequency of 620 kHz. Unfortunately, this situation is not so easy to attain at the higher frequencies used by many communication receivers. In these cases, a technique known as *double conversion* is employed to solve the image frequency problems. This process will be described in Chapter 7.

The use of an RF amplifier with its own input tuned circuit also helps to minimize this problem. Now the image frequency must pass through two tuned circuits (that are tuned to a different frequency) before it is mixed. These tuned circuits at the input of the RF and mixer stages obviously serve to attenuate the image frequency to a greater extent than can the single tuned circuit in receivers without an RF stage.

RF Amplifiers

The use of RF amplifiers in superheterodyne receivers varies from none in undemanding applications up to three or even four stages in sophisticated communication receivers. Even inexpensive AM broadcast receivers that contain no RF amplifier do contain an RF section—the tuned circuit at the mixer's input. The major benefits of using RF amplification are the following:

1. Improved image frequency rejection.
2. More gain and thus better sensitivity.
3. Improved noise characteristics.

The first two are self-explanatory at this point, but the last advantage requires further elaboration. Mixer stages require devices to be operated in a nonlinear area in order to generate the required difference frequency at their output, the IF input. This process is inherently more noisy (i.e., higher NF) than normal class A linear bias. Thus, the use of RF amplification stages to bring up the signal to appreciable levels makes the mixer noise effect less noticeable.

The RF amplifier usually employs a FET as its active component. While BJTs certainly can be utilized, the following advantages of FETs have led to their general usage in RF amplifiers.

1. Their high input impedance does not "load" down the Q of the tuned circuit preceding the FET. It thus serves to keep the selectivity at the highest possible level.
2. The availability of dual-gate FETs allows for simple injection of the AGC signal.
3. Their input/output square-law relationship allows for lower distortion levels.

The distortion referred to in the last item is called *cross-modulation* and is explained in Chapter 6.

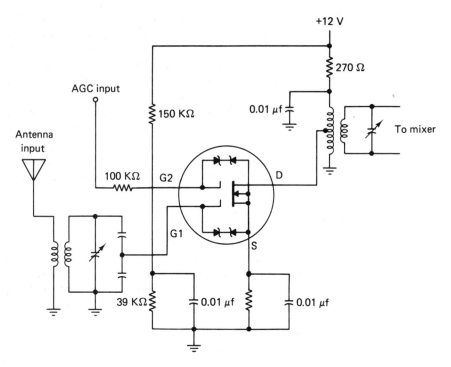

Fig. 3-20. Dual-gate MOSFET RF amplifier.

A typical MOSFET RF amplifier stage is shown in Fig. 3-20. It is a dual-gate unit, with the AGC level applied to gate 2 to provide for automatically variable gain. The received antenna signal is fed via a tuned coupling circuit to gate 1. The gate 1 and output drain connections are tapped down on their respective coupling networks, which keeps the device from self-oscillation without the need for a neutralizing capacitor. Notice the built-in transient protection shown within the symbol for the 40673 MOSFET. Zener diodes between the gates and source/ substrate connections provide protection from up to 10-V p-p transient voltages. This is a valuable safeguard because of the extreme fragility of the MOSFET gate/ channel junction.

Mixer/LO

The frequency conversion accomplished by the mixer/LO combination can be accomplished in a number of ways. The circuits shown in Fig. 3-21 illustrate some of the possiblilities. They all make use of a device's nonlinearity to generate sum and difference frequencies between the RF and local oscillator signal to generate output components at the IF frequency. The circuit at *A* in Fig. 3-21 utilizes a BJT and requires a separate oscillator (quite often a Hartley version) input, as do the circuits at *B* and *C*, which use a JFET and MOSFET as the nonlinear element required to generate sum and difference frequencies. The circuit shown at

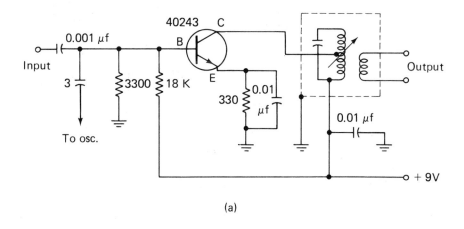

(a)

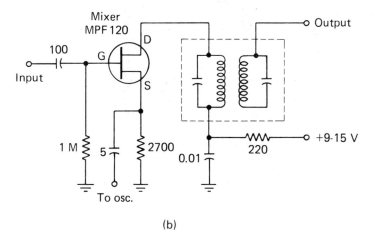

(b)

Fig. 3-21. Typical mixer circuits.

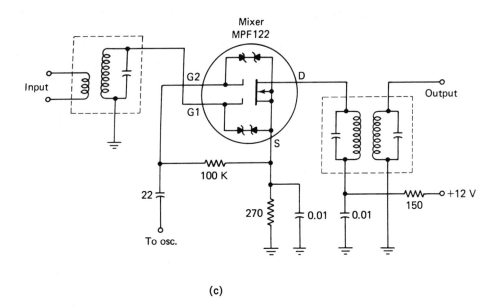

(c)

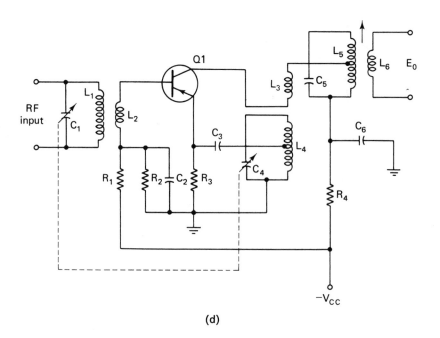

(d)

Fig. 3-21. (Cont.).

D is a *self-excited mixer* in that a single device does the mixing and generates the LO frequency. Self-excited mixers are sometimes referred to as *autodyne mixers*. The oscillator-tuned circuit of C_4 and L_4 provides a positive feedback signal to maintain oscillation via coil L_3, which is magnetically coupled to L_4. The oscillator signal is injected into Q_1's emitter via C_3 and the RF signal into its base via L_1–L_2 transformer action. The "mixed" output at Q_1's collector is fed to the C_5–L_5 tank circuit, which "tunes" in the desired frequency for the IF amplifiers. Recall that mixing signals through a nonlinear device generates many frequency components; the tuned circuit is used to select the desired ones for the IF amplifiers.

A diode could be used as the mixer element but seldom is because the use of transistors provides additional gain as well as frequency conversion. Mixers are also referred to as *converters* and the *first detector*.

IF Amplifiers

The IF amplifiers provide the bulk of a receiver's gain (and thus are a major influence on its sensitivity) and selectivity characteristics. An IF amplifier is not a whole lot different than an RF stage except it operates at a fixed frequency. This allows the use of fixed double-tuned inductively coupled circuits to allow for the sharply defined bandpass response characteristic of superheterodyne receivers.

The number of IF stages in any given receiver varies, but from two to four is typical. Some typical IF amplifiers are shown in Fig. 3-22. The circuit at *A* uses the

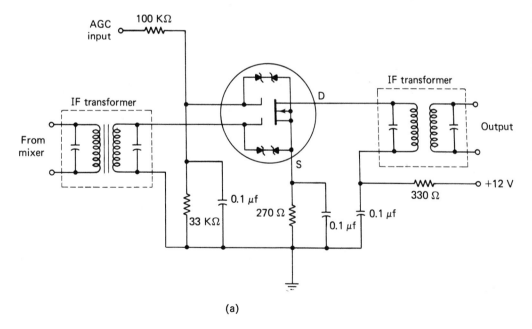

(a)

Fig. 3-22. Typical IF amplifiers.

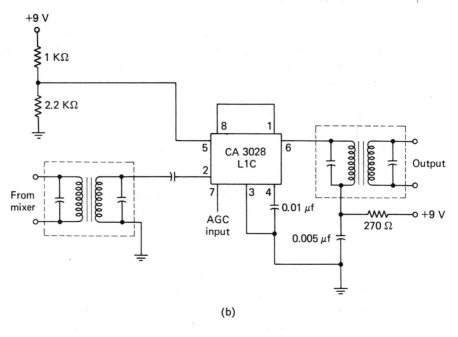

(b)

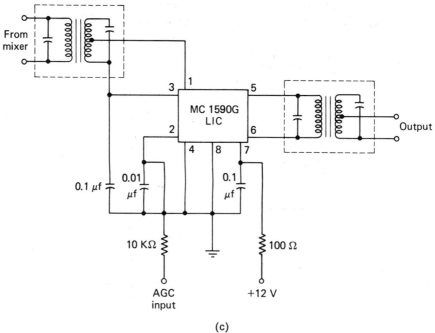

(c)

Fig. 3-22. (Cont.).

40673 dual-gate MOSFET while the other two use LICs specially made for IF amplifier applications. Notice the double-tuned LC circuits at the input and output of all three circuits. They are shown within dotted lines to indicate they are one complete assembly. They can be economically purchased for all common IF frequencies and have a variable slug in the transformer core for fine tuning their center frequency. All three of the circuits have provision for the AGC level. Note that not all receiver's utilize AGC to control the gain of mixer and/or RF stages, but they invariably do control the gain of the IF stages.

3-6 AGC

The purpose of automatic gain control (AGC) has already been explained. Without this function, a receiver's usefulness is seriously impaired. The following list details some of the problems that would be encountered in a receiver without this provision:

1. Tuning the receiver would be a nightmare. In order to be sure not to miss the weak stations, you would have the volume control (in the non-AGC set) turned way up. As you tuned into a strong station, you would probably blow out your speaker and/or eardrums.
2. The received signal from any given station is constantly changing as a result of changing weather and ionospheric conditions. The AGC allows you to listen to a station without constantly monitoring the volume control.
3. Many radio receivers are utilized under mobile conditions. For instance, a standard broadcast AM car radio would be virtually unusable without a good AGC to compensate for the change in signal in different locations.

Obtaining the AGC Level

Most AGC systems obtain the AGC level just following the detector. Recall that following the detector diode, an *RC* filter removes the high frequency but hopefully leaves the low-frequency envelope intact. By simply increasing that *RC* time constant a slowly varying dc level is obtained. The dc level changes with variations in the strength of the overall received signal.

A in Fig. 3-23 shows the output from a diode detector with no filtering. In this case, the output is simply the AM waveform with the positive portion rectified out for two different levels of received signal into the diode. At *B* the addition of a filter has provided the two different envelope levels while filtering out the high-frequency content. These signals correspond to an undesired change in volume of

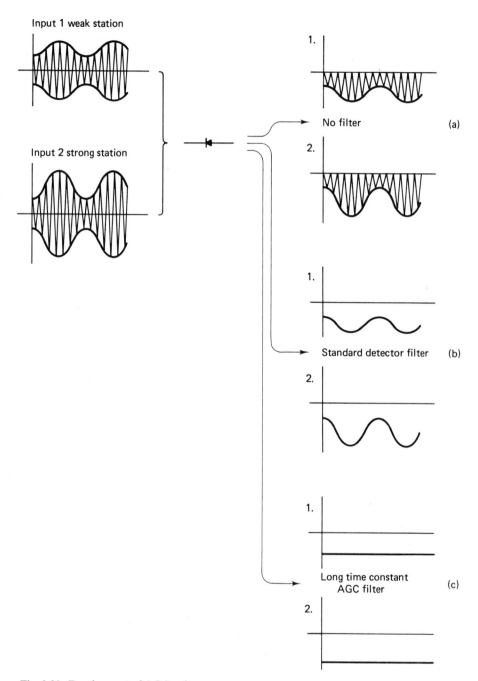

Input 1 weak station

Input 2 strong station

1.

No filter   (a)

2.

1.

Standard detector filter   (b)

2.

1.

Long time constant AGC filter   (c)

2.

Fig. 3-23. Development of AGC voltage.

two different received stations. At *C*, a much longer time constant filter has actually filtered the output into a dc level. Notice that the dc level changes, however, with the two different levels of input signal. This is a typical AGC level that is subsequently fed back to control the gain of previous IF stages and/or the mixer and RF stages.

In this case, the larger negative dc level at C_2 would cause the receiver's gain to be decreased such that the ultimate speaker output is roughly the same for either the weak or strong station. It is important that the AGC time constant be long enough so that desired radio signal level changes that constantly occur do *not* cause a change in receiver gain. The AGC should only respond to average signal strength changes and as such usually has a time constant of about a full second.

Controlling the Gain of a Transistor

Figure 3-24 illustrates a method whereby the variable dc AGC level can be used to control the gain of a common emitter (CE) transistor amplifier stage. In the case of a strong received station, the AGC voltage developed across the AGC filter capacitor (C_{AGC}) is a large negative value which subsequently lowers the forward bias on Q_1. It causes more dc current to be drawn through R_2, and hence less

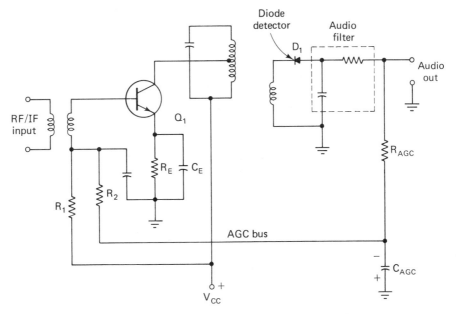

Fig. 3-24. AGC circuit illustration.

is available for the base of Q_1, since R_1, which supplies current for both, can only supply a relatively constant amount. The voltage gain of a CE stage with an emitter bypass capacitor (C_E) is nearly directly proportional to dc bias current, and therefore the strong station reduces the gain of Q_1. Subsequently, the reception of very weak stations would reduce the gain of Q_1 very slightly, if at all. The introduction of AGC back in the 1920s marked the first major use of an electronic feedback control system. The AGC feedback path is called the AGC bus, because in a full receiver it is usually "bussed" back into a number of stages to obtain a large amount of gain control. Some receivers require more elaborate AGC schemes and they will be examined in Chapter 8.

3-7 AM RECEIVER SYSTEMS

We have thus far examined the various sections of AM receivers. It is now time to "put it all together" and look at the complete system. Figure 3-25 shows the schematic of a low-cost, widely used circuit for a low-cost AM receiver. It is the basic configuration found in virtually all pocket AM transistor receivers. These six transistor units can usually be purchased for less than $5, an amazingly low price made possible by the extremely large quantities of production. When first introduced in the late 1950s, however, they cost in the vicinity of $50, primarily because of the initial engineering costs and the higher prices of semiconductors. The schematic shown in Fig. 3-25 shows only four of the six transistors, since the push-pull audio power amp, which requires two more transistors, has been omitted.

The L_1, L_2 inductor combination is wound on a powdered-iron (ferrite) core and functions as an antenna as well as an input coupling stage. Ferrite core loopstick antennas offer extremely good signal pick-up, considering their small size, and are adequate for the strong signal strengths found in urban areas. The RF signal is then fed into Q_1, which functions as the mixer and local oscillator (self-excited). The ganged tuning capacitor, C_1, tunes to the desired incoming station (the B section) and adjusts the LO (the D section) to its appropriate frequency. The output of Q_1 contains the IF components, which are tuned and coupled to Q_2 by the T_1 package. The IF amplification of Q_2 is coupled via the T_2 IF "can" to the second IF stage, Q_3, whose output is subsequently coupled via T_3 to the diode detector E_2. Of course, T_1, T_2, and T_3 are all providing the very good superheterodyne selectivity characteristics at the standard 455 kHz IF frequency. The E_2 detector diode's output is filtered by C_{11} such that just the intelligence envelope is fed via the R_{12} volume control potentiometer into the Q_4 audio amplifier. The AGC filter, C_{12}, then allows for a fed-back control level into the base of Q_2.

This receiver also illustrates the use of an *auxiliary AGC diode* (E_1). Under normal signal conditions, E_1 is reverse-biased and has no effect on the operation. At some predetermined high signal level, the regular AGC action causes the dc

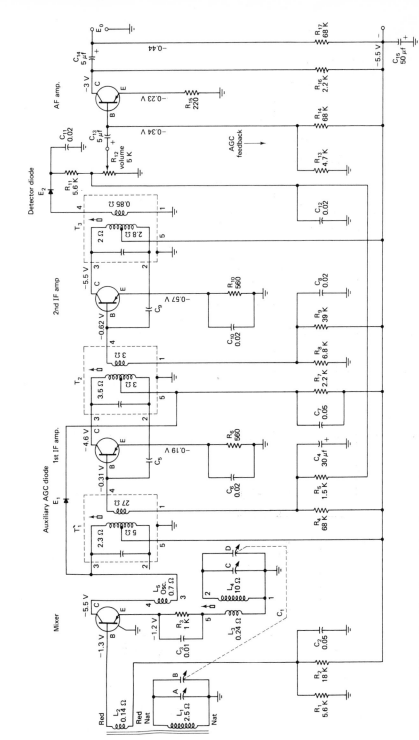

Fig. 3-25. AM broadcast band superheterodyne receiver.

95

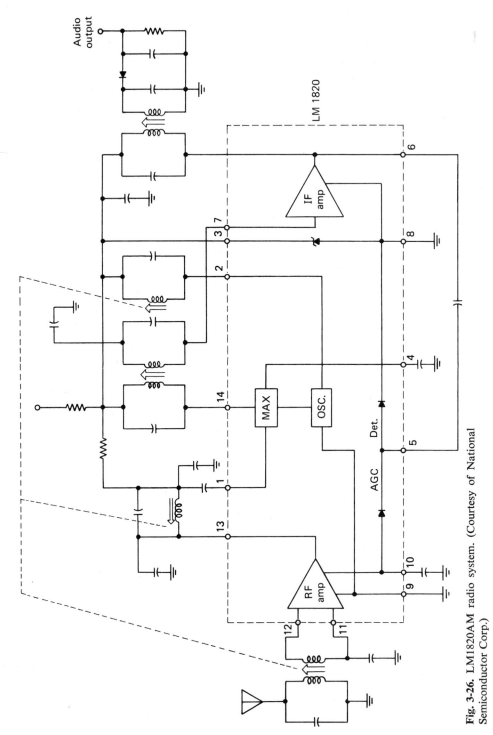

Fig. 3-26. LM1820AM radio system. (Courtesy of National Semiconductor Corp.)

level at E_1's cathode to decrease to the point where E_1 starts to conduct (forward bias) and it loads down the T_1 tank circuit, thus reducing the signal coupled into Q_2. The auxiliary AGC diode thus furnishes additional gain control for strong signals and enhances the range of signals that can be compensated for by the receiver.

LIC AM Receivers

The complete function of a superheterodyne AM receiver can be accomplished with special LICs. The only hitch is that the tuned circuits must be added on externally due to the limitations of ICs in this area. A number of AM chips are available from the various IC manufacturers. The RCA CA3088E is a typical unit and is shown schematically in Fig. 3-26(b), and a block diagram with the associated external tuning circuitry is shown at *C*. A suggested configuration for applications that require an RF amplifier (such as car radios) is shown at *D*.

Even though the use of the LIC greatly reduces component count, the physical size and cost are not appreciably affected, since they are mainly determined by the frequency selective circuits. Thus, LIC AM radios are not widely used for low-cost applications but do find their way into higher-quality AM receivers, where certain performance and feature advantages can be realized.

The limiting factor of tuned circuits is the only roadblock to having complete receivers on a chip except for the station selection and volume controls. Alternatives to *LC*-tuned circuits, such as ceramic filters, may in the future be integrable (see Chapter 8 for further detail on alternative filter circuits). Another possibility is the use of phase-locked loop (PLL) technology in providing a nonsuperheterodyne type of receiver. (See Chapter 6 for a more general discussion of PLL theory.) Using this approach, it is theoretically possible to fabricate a functional AM broadcast band receiver using just the chip and two external potentiometers (for volume control and station selection) and the antenna. Figure 3-27 shows the basics of this system utilizing the 561PLL. There is no obvious reason why the external components as well as an RF amplifier and the audio amplifiers could not be incorporated into a single chip, but future research and engineering effort on this approach will dictate its ultimate feasibility.

The PLL is locked to the incoming AM carrier with the voltage-controlled oscillator (VCO) providing the LO signal. The amplitude of the detected intelligence at pin 1 is a function of the phase relationship between the AM signal and the LO. The C_y, R_y combination provides the proper phase shift for the standard AM band with values of 135 pF and 3 kΩ, respectively.

The low-pass filter for the loop, C_L, is not critical, since no information is being derived from the loop error. A value of 10 nF is suitable and is only to ensure loop stability. To tune, the VCO is set to the station frequency. A value of 330 pF for C_o sets the VCO up to oscillate at 940 kHz, while the 5-kΩ potentiometer allows a fine adjust to cover the complete AM band on either side of 940 kHz.

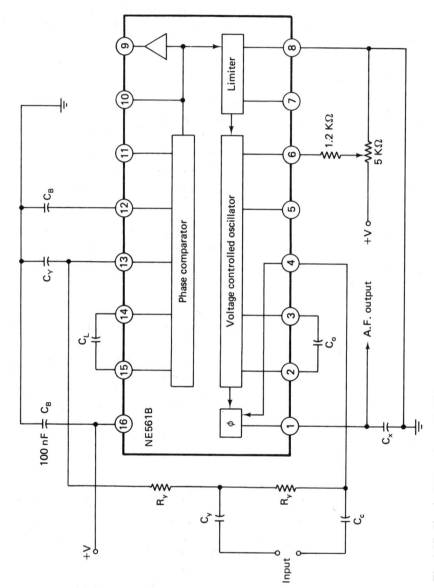

Fig. 3-27. Simple AM receiver with PLL.

The choice of C_x is determined by the desired audio bandwidth. A cutoff frequency of 5 kHz is therefore suitable for the broadcast AM band. Its value should be calculated in conjunction with the 8-kΩ resistance seen looking into pin 1 of the chip, in parallel with the input impedance of whatever subsequent audio amplifier is used.

4

SINGLE-SIDE-BAND COMMUNICATIONS

4-1 SINGLE-SIDE-BAND CHARACTERISTICS

The basic concept of single-side-band (SSB) communications was understood as early as 1914. It was first realized through mathematical analysis of an amplitude-modulated RF carrier. It has been shown that a carrier amplitude-modulated by a single sine wave of voltage consists of three different frequencies: (1) the original carrier with amplitude unchanged; (2) a frequency equal to the difference between the carrier and the modulating frequencies, with an amplitude equal to a maximum of one-half that of the modulating signal; and (3) a frequency equal to the sum of the carrier and the modulating frequencies, with an amplitude also equal to a maximum of one-half that of the modulating signal. The two new frequencies of course are the side frequencies.

Upon recognition of the fact that side bands existed, further studies were made. These investigations and experiments proved that after the carrier and one of the side bands were eliminated, the other side band could be used to transmit the intelligence. (Since the carrier amplitude and frequency never change, there is no intelligence contained in the carrier.) Further experiments proved that both side bands could be transmitted, each containing different intelligence, with a suppressed or completely eliminated carrier.

By 1923, the first patent for this system had been granted, and a successful SSB communications system was established between the United States and England. Today, SSB communication plays an increasingly vital role in radio communications because of its many advantages over standard AM systems. The FCC, recognizing these advantages, further increased its use by requiring most transmissions in the overcrowded 2- to 30-MHz range to be SSB starting in 1977.

Power Distribution

In the composite AM waveform, there are two values of power to be considered: the carrier power and the peak power. *Carrier power* is the average power in the unmodulated carrier. *Peak power* is the power developed when the carrier and both side bands are in phase. If we know the voltages, we can calculate the power in any of the components of the AM waveform.

If we have a carrier modulated 100% by a single sine wave of voltage of constant amplitude, the computations are not too difficult. Assume that the rms voltage of the unmodulated carrier is 100 V and the antenna radiation resistance is 50 Ω. The power in the unmodulated carrier can be found from the power formula

$$P = \frac{E^2}{R}$$

$$= \frac{(100)^2}{50} = \frac{10,000}{50} = 200 \text{ W}$$

Since the amplitude of the carrier remains constant, modulated or unmodulated, the power in the carrier is 200 W at all times.

For 100% modulation, the modulating signal voltage is equal to the carrier voltage, in this case 100 V. The rms value of each sideband voltage is 50 V. The power in either side band can be found as

$$P = \frac{(50)^2}{50} = 50 \text{ W}$$

As you can see, the power in each side band is one-fourth the power in the carrier at 100% modulation. The total side-band power is 100 W. With a carrier power of 200 W, this gives a total power of 300 W. This means that to attain 100% modulation of an RF carrier with a sine-wave modulating signal, a modulating power equal to one-half the carrier power is required. The additional modulating power is divided equally between the upper and lower side bands.

The peak power for this 100% modulated wave can be found as follows. The rms value of $E_{\max}$ is 200 V. The 100-V carrier increases to 200 V and decreases to 0 V during 100% modulation. Therefore,

$$P = \frac{(200)^2}{50} = \frac{40,000}{50} = 800 \text{ W (peak)}$$

While all the intelligence is contained in the side bands, two-thirds of the total power is in the carrier. It would appear that a great amount of power is wasted during transmission. The basic principle of single-side-band transmission is to eliminate or greatly suppress the high-energy RF carrier. This can be accomplished without affecting the fidelity of the emitted intelligence, since the carrier contains no intelligence.

If a means of suppressing or completely eliminating the carrier is devised, the power that was used for the carrier can be converted into useful power to transmit the intelligence in the side bands. Since both upper and lower side bands contain the same intelligence, one of these could also be eliminated, thereby cutting the bandwidth required for transmission in half.

Power Comparison

The total power output of a conventional AM transmitter is equal to the carrier power plus the power output of the side bands. Conventional AM transmitters are rated in carrier power output. In single-side-band operation, there is either a greatly attenuated carrier or a completely eliminated carrier. For this reason, the SSB transmitter is often rated in peak envelope power (pep). In the example of the 200-W carrier transmitter previously used, the peak power was 800 W. If this same transmitter were converted to SSB operation, it would be rated as an 800-W single-side-band transmitter, and a 6-dB gain in power would be realized. (Each doubling of power equals a 3-dB gain.) This gain is due to the elimination of the carrier and one of the side bands.

The bandwidth of the 800-W SSB transmitter would be only one-half that required for a conventional AM transmitter. Using only half the BW means that the external noise picked up by the receiver is cut in half. This is an advantage that results in about a 3-dB gain at the receiver. This gain is due to the better signal-to-noise ratio obtained.

Types of Side-Band Transmission

Several types of single-side-band systems have been developed. Therefore, it is necessary to clarify the various applications of the transmission principle involved. The term *single-side-band* (SSB) *modulation* is defined as a form of amplitude modulation in which one side band and the carrier are suppressed.

In one system, the carrier and one of the side bands are completely eliminated at the transmitter; only one side band is transmitted. This is standard single side band, or simply SSB; it is quite popular with amateur radio operators. The chief advantage of this system is maximum transmitted signal range with minimum transmitter power.

Another system eliminates one side band and suppresses the carrier to a desired level. The suppressed carrier is then used at the receiver for a reference, AGC, automatic frequency control (AFC), and, in some cases, demodulation of the intelligence-bearing side band. This is called a single-side-band suppressed carrier (SSSC).

The type of system most commonly used in military communications is referred to as *twin-side-band suppressed carrier*, or *independent side-band* (ISB) *transmission*. This system involves the transmission of two independent sidebands, each containing different intelligence, with the carrier suppressed to a desired level.

Advantages of SSB

The most important advantage of SSB systems is a more effective utilization of the available frequency spectrum. The bandwidth required for the transmission of one conventional AM signal contains two equivalent SSB transmissions. This type of communications is, therefore, especially adaptable to the already over-crowded high-frequency spectrum.

A second advantage of this system is that it is less subject to the effects of selective fading. This is because there is *no* definite phase relationship between the upper and lower side bands and the carrier as there is in conventional AM. In the propagation of conventional AM transmissions, if the upper-side-band frequency strikes the ionosphere and is refracted back to earth at a different angle from that of the carrier and lower-side-band frequencies, distortion is introduced at the receiver. Under extremely bad conditions, complete signal cancellation may result, which means a complete loss of the intelligence. The two side bands should be identical in phase so that when passed through a nonlinear device (i.e., a diode detector), the difference between the side bands and carrier is identical. That difference is the intelligence and will be distorted if the two side bands have a phase difference.

Another advantage realized by all types of SSB systems is that a higher percentage of power is in the radiated intelligence. The SSB system has an overall 9 dB gain, 6 dB at the transmitter and 3 dB at the receiver, as previously explained as compared to AM. This, in effect, means that a SSB transmission of 100 W is equivalent in range to a much-higher-powered AM transmission of about 800 W This, coupled with the fact that the SSB transmission requires half the BW of AM, explains the surging use of SSB.

Disadvantages of SSB

Obviously, the use of SSB has some disadvantages. One of the disadvantages of SSB equipment is its high cost. This results from the complex circuitry and type of components used. Special emphasis is placed upon voltage regulation, stability, and reliability. The oscillator frequencies involved are very critical and must be stabilized; otherwise, distortion will occur. In SSB systems, every precaution is taken to prevent frequency drift, because when the carrier drifts, the side bands will drift proportionately. If corrections are not made, the intelligence will be distorted upon demodulation.

4-2 SIDE-BAND GENERATION: THE BALANCED MODULATOR

Before single-side-band suppressed carrier transmission can occur, two things must be accomplished: (1) the carrier must be eliminated, or suppressed; and (2) one of the two side bands must be filtered out. Once this has been achieved, the selected side band can be applied to the final power amplifier for transmission. A *balanced modulator* circuit can be used to generate only the two side bands. Balanced modulation is a system of adding intelligence to a carrier wave whereby only the side bands are produced; the carrier wave is eliminated.

The balanced modulator resembles the conventional push-pull amplifier in circuitry but not in operation. Figure 4-1 shows a conventional push-pull amplifier.

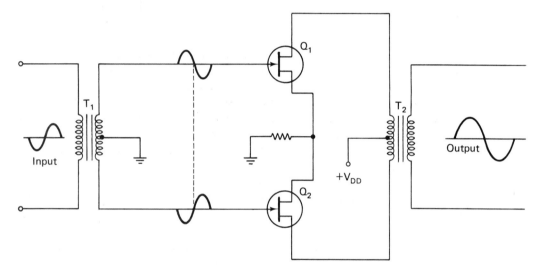

Fig. 4-1. Conventional push-pull amplifier.

For proper operation as an amplifier, the signals applied to the gates of Q_1 and Q_2 must be 180° out of phase. When the positive cycle is applied to the gate of Q_1, drain current flowing through T_2 will produce the positive portion of the output voltage. When the gate of Q_1 goes negative, the gate of Q_2 goes positive and drain current from Q_2 flows in the opposite direction through the primary of T_2, producing the negative-going alternation of the output waveshape.

By slightly modifying the push-pull amplifier, as shown in Fig. 4-2, no output signal will be obtained when an input signal is applied. Notice that the input signal applied to the gates of Q_1 and Q_2 in Fig. 4-2 has the same polarity. Both gates are positive at the same time. A positive-going gate in each FET causes the drain current to increase simultaneously in each FET. Since the drain current of Q_1 is flowing down through the primary of T_2 at the time the drain current of Q_2 is flowing up, the two magnetic fields produced in T_2 will effectively cancel. Thus, no output signal will be developed. At first glance, this particular circuit is not per-

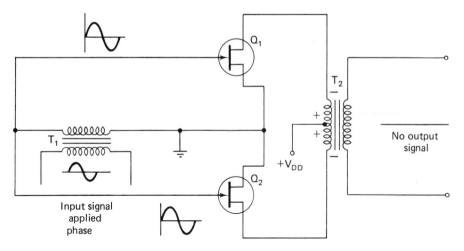

Fig. 4-2. Push-pull amplifier with signal applied in phase.

forming any useful function. On the other hand, assume the input signal to be the carrier frequency. When the carrier frequency is applied in phase, no output signal will be developed in the output; thus, the carrier frequency has been suppressed, or eliminated, in the output.

Again, modify the push-pull amplifier circuit and get the circuit shown in Fig. 4-3 (essentially a combination of Figs. 4-1 and 4-2). In this simplified push-pull balanced modulator, the carrier input is applied to the gates of both Q_1 and Q_2 in phase; when the RF voltage is going positive, both gates are driven positive. The modulating voltage is applied to the two gates in the conventional push-pull manner; when the modulating signal drives the gate of Q_1 in a positive direction, the gate of Q_2 is driven in a negative direction.

Now consider the effect of the RF voltage on the stage. Since the carrier frequency causes both gates to be driven positive at the same time, the drain current of both Q_1 and Q_2 increases at the same time. Consequently, the magnetic fields caused by the two equal drain currents are effectively canceled in the output transformer T_2, which is center-tapped to $+V_{DD}$. On the negative swing of the RF voltage, the drain current of both FETs decreases at the same time. Again, the magnetic fields cancel. The carrier frequency, therefore, does not appear in the output.

The effect of the modulating voltage is different. When the positive swing of the modulating voltage is applied, the gate of Q_2 is driven negative. On the negative alternation of the modulating voltage, the gate of Q_1 is driven negative, but the gate of Q_2 is driven positive. Because of this push-pull arrangement, drain current flows on both the positive and negative swings, first through one FET and then through the other. There is no cancellation in the output.

When the modulating and RF voltages are applied at the same time, first one FET conducts heavily and then the other, depending upon which gate is driven

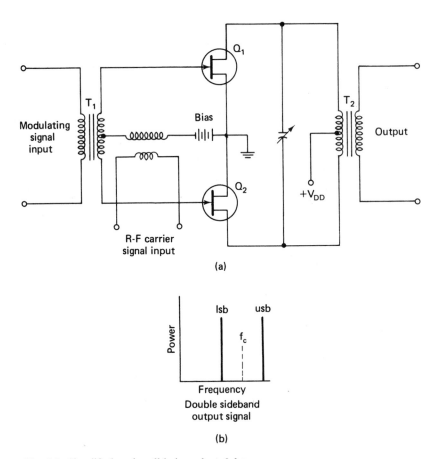

Fig. 4-3. Simplified push-pull balanced modulator.

positive by the modulating voltage. The modulating voltage acts as a varying bias. Assuming nonlinear operation, when the gate of Q_1 is positive, RF currents at the carrier frequency, the modulating frequency, and the sum and difference frequencies flow through Q_1. The carrier frequency currents are effectively canceled by the carrier frequency currents from Q_2 in the common drain load, as explained before. The sum and difference frequencies cause induced voltages in the secondary of transformer T_2, and the modulating frequency is so low that it is highly attenuated at the output of T_2.

The output signal of the balanced modulator thus consists of the upper and lower side bands; the carrier and modulating frequencies have been eliminated. This is shown in Fig. 4-3(b). As in AM generation, the generation of sum and difference frequencies relies on nonlinear operation. A tuned circuit is required to select the desired output frequencies from the other components generated by the heterodyning process.

Balanced Ring Modulator

The other often used means of eliminating the carrier is called the *balanced ring modulator*. Figure 4-4 shows a ring modulator schematically. Considering just the carrier with the instantaneous polarity as shown, electron flow is as indicated by the arrows. The current flow through both halves of L_5 are equal but opposite, and thus the carrier is canceled in the output. This is also true on the carrier's other half-cycle, only now diodes B and C conduct instead of A and D.

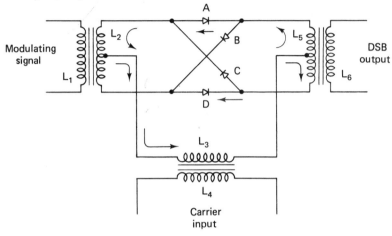

Fig. 4-4. Balanced ring modulator.

Considering just the modulating signal, current flow occurs from winding L_2 through diodes C and D or A and B but not through L_5. Thus, there is no output of the modulating signal either. Now with both signals applied, but with the carrier amplitude much greater than the modulating signal, the conduction is determined by the polarity of the carrier. The modulating signal either aids or opposes this conduction. With the modulating signal polarity as shown in Fig. 4-4, diode D will conduct more than A and the current balance in winding L_5 is upset, which causes outputs of the desired side bands but continued suppression of the carrier. This modulator is capable of 60 dB of carrier suppression when carefully matched diodes are utilized. It relies on the nonlinearity of the diodes to generate the sum and difference frequencies.

LIC Balanced Modulator

A balanced modulator of either type previously explained requires extremely well matched components to provide good suppression of the carrier (40 or 50 dB suppression is usually adequate). This suggests the use of LICs because of the superior component matching characteristics obtainable when devices are fabricated on the same silicon chip. A number of devices specially formulated for balanced modulator applications are available. A data sheet for a representative sample, the Plessey Semiconductors SL640C/641C, is shown in Fig. 4-5. In essence,

SL640C & SL641C
DOUBLE BALANCED MODULATORS

The SL640C is designed to replace the conventional diode ring modulator, in RF and other communications systems, at frequencies of up to 75MHz. It offers a performance competitive with that of the diode ring while eliminating the associated transformers and heavy carrier drive power requirements.

At 30 MHz, carrier and signal leaks are typically −40dB referred to the desired output product frequency. Intermodulation products are −45dB with a 60 mV rms input signal

The SL641C is a version of the SL640C intended primarily for use in receiver mixer applications for which it offers a lower noise figure and lower power consumption. No output load resistor is included and signal leakage is higher, but otherwise the performance is identical to that of the SL640C

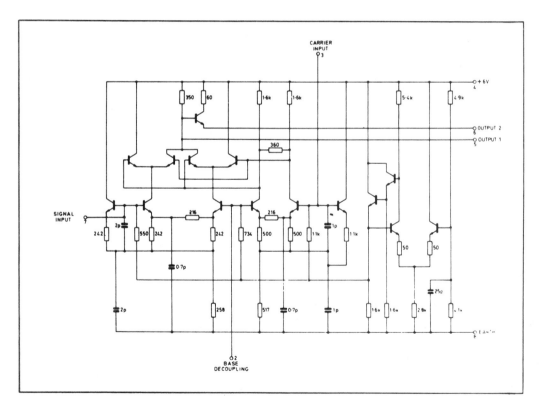

Fig. 4-5. Balanced modulator LIC. (Courtesy of Plessey Semiconductor.)

ELECTRICAL CHARACTERISTICS SL640C & SL641C

Test conditions: Supply voltage = +6V
Temperature = +25°C unless otherwise stated

Characteristic	Circuit	Value			Units	Test Conditions
		Min.	Typ.	Max.		
Conversion gain	SL640C	−2	0	+2	dB	Signal: 70mVrms, 1.75MHz Carrier: 100mVrms, 28.25MHz Output: 30MHz
Signal leak = [Signal output / Desired sideband output]	SL640C		−40	−20	dB	
Carrier leak = [Carrier output / Desired sideband output]	SL640C		−40	−20	dB	
Intermodulation products	SL640C		−45	−35	dB	Signal 1: 42.5mVrms, 1.75MHz Signal 2: 42.5mVrms, 2MHz Carrier: 100mVrms, 28.25MHz Output: 29.75MHz
Conversion transconductance	SL641C	2.2	2.5	3.5	mmho	Signal: 70mVrms, 30MHz Carrier: 100mVrms, 28.25MHz Output: 1.75MHz
Signal leak	SL641C		−18	−12	dB	
Carrier leak	SL641C		−25	−12	dB	
Intermodulation products	SL641C		−45	−30	dB	Signal 1: 42.5mVrms, 30MHz Signal 2: 42.5mVrms, 31MHz Carrier: 100mVrms, 28.25MHz Output: 3.75MHz
Carrier input impedance	Both		1kΩ & 4pF			
Signal input impedance	SL640C SL641C		500Ω & 5pF 1kΩ & 4pF			
Output impedance (see Operating Notes)	SL640C SL641C		350Ω & 8pF 8		pF	Output 1
Max. input before limiting	SL640C SL641C		210 250		mVrms mVrms	
Quiescent current consumption	SL640C SL641C		12 10	16 13	mA mA	
Noise figure	SL640C SL641C		15 12		dB dB	
Signal leak variation	Both		±2		dB	−55°C to +125°C
Carrier leak variation	Both		±2		dB	
Conversion gain variation	Both		±1		dB	

ABSOLUTE MAXIMUM RATINGS

Storage temperature range	−55°C to +175°C
Chip operating temperature	+175°C
Chip-to-ambient thermal resistance	250°C/W
Chip-to-case thermal resistance	80°C/W
Supply voltage	+9V
Free air operating temperature range	−55°C to +125°C

Fig. 4-5. (Cont.).

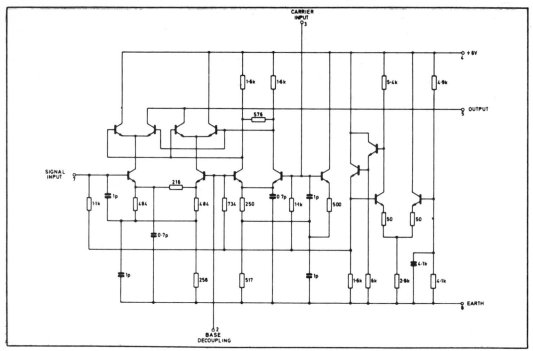

Fig. 2 Circuit diagram of SL641C

OPERATING NOTES

The SL640C circuit requires input and output coupling capacitors which normally should be chosen to present a low reactance compared with the input and output impedances (see electrical characteristics). However, for minimum carrier leak at high frequencies the signal input should be driven from a low impedance source, in which case the signal input capacitor reactance should be comparable with the source impedance.

Pin 2 must be decoupled to earth via a capacitor which presents the lowest possible impedance at both carrier and signal frequencies. The presence of these frequencies at pin 2 would give rise to poor rejection figures and to distortion.

If the emitter follower is used, an external load resistor must be provided to supply emitter current. The quiescent output voltage from the emitter follower (pin 6) is +4.6V. To achieve maximum rejection figures at high frequencies, pin 1 (which is connected to the header) should be connected to earth and effective HT decoupling should be employed. The DC impedance should not exceed 800 ohms.

The SL641C is very similar to the SL640C but has, instead of a voltage output, a current output to enable a tuned circuit to be directly connected.

If both output sidebands are developed across the load (i.e. wideband operation), the AC impedance of the load must be less than 800Ω. If the output at one sideband frequency is negligible, the AC impedance may be raised to 1.6kΩ. It may be further raised if it is not desired to use the maximum input swing of 210mV rms.

The SL640C/641C may be used with supply voltages of up to +9 volts with increased dissipation.

Signal and carrier leaks may be minimised with 10kΩ potentiometers and 330kΩ resistors connected as shown in fig.3. R1 is adjusted to minimise signal leak; R2 to minimise carrier leak.

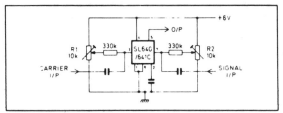

Fig. 4-5. (Cont.).

Fig. 3 Signal and carrier leak adjustments

all that is required is the device, a coupling capacitor for both the carrier and intelligence inputs, and a dc supply. This device does not require the transformers used in the previously explained balanced modulators.

4-3 SSB FILTERS

Once the carrier has been eliminated, it is necessary to cancel one of the side bands without affecting the other one. This requires a sharply defined filter, as Fig. 4-6 helps illustrate. Voice transmission requires audio frequencies from about 100 Hz to 3 kHz. Therefore, the upper and lower side bands generated by the balanced modulator are separated by 200 Hz, as shown in Fig. 4-6. For a 10-MHz carrier, this implies a filter Q of

$$Q = \frac{10 \text{ MHz}}{200 \text{ Hz}} = 50,000$$

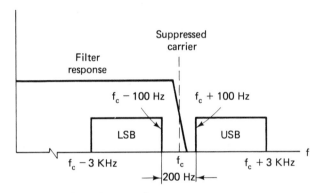

Fig. 4-6. Sideband suppression.

This extremely high Q requirement is usually circumvented by generating the side bands around a low carrier frequency such as 100 kHz, which lowers the Q required to

$$Q = \frac{100 \text{ kHz}}{200 \text{ Hz}} = 500$$

Then, after removing one side band, an additional frequency translation is usually employed to get the side band up to the desired frequency range.

Both SSB transmitters and receivers require extremely sensitive bandpass filters in the region of 100 to 500 kHz. In receivers a high order of adjacent channel rejection is required if channels are to be closely spaced in order to conserve spectrum space. In SSB transmitters the signal bandwidth must be limited sharply in order to pass the desired side band and reject the other. The filter used, there-fore, must have very steep skirt characteristics (fast cutoff) and a flat bandpass

characteristic in order to pass all frequencies in the band equally well. These
filter requirements are met by crystal filters, ceramic filters, *LC* filters, and mechanical filters.

Simple Filters

Filters are designed to produce high attenuation at particular frequencies.
Filters are made up of inductors, capacitors, and resistors, arranged to offer a high
loss to signals of undesired frequencies and a low loss to other frequencies. There
are three basic types of filters: the low-pass filter, the high-pass filter, and the
bandpass filter. They can be in *RC*, *LR*, or *LC* configurations. However, the use
of *LC* configurations offers sharper pass bands as are required for the high *Q*s
of SSB applications.

A simple low-pass filter is shown in Fig. 4-7(a). When low-frequency signals
are applied to a low-pass filter, the filter causes very little loss. At low frequencies
the reactance of the series inductors is low and the reactance of the shunt capacitor
is high. However, when high-frequency signals are applied to a low-pass filter, the
reverse is true; the reactance of the series inductors is high, and the reactance of the
shunt capacitor is low. The frequency response of a low-pass filter is illustrated
below the schematic diagram; f_{hc} is the high frequency cutoff and it is the point
where 3 dB attenuation occurs compared to the pass band.

In a high pass filter, the capacitors are in the series arm and the inductor is
in the shunt arm. A high-pass filter offers a high loss to low-frequency signals

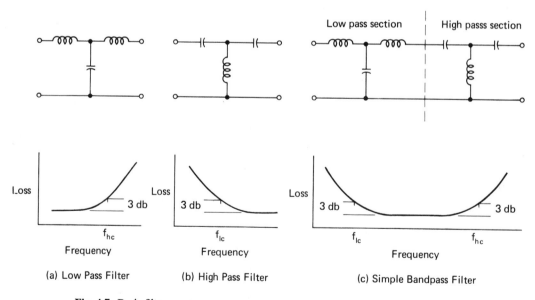

(a) Low Pass Filter   (b) High Pass Filter   (c) Simple Bandpass Filter

Fig. 4-7. Basic filters.

and a low loss to high-frequency signals. The relation between output/input and frequency is shown in Fig. 4-7(b). In this illustration, f_{lc} is the low-frequency cutoff.

A simple bandpass filter can be made from a combination of these low- and high-pass filters, as shown in Fig. 4-7(c). Assume that the low-pass section (coils in the series arm) is designed to pass a frequency range from zero to 2000 Hz and to suppress frequencies above that point. In addition, assume that the high-pass section (series capacitors) is designed to pass a frequency range from 1000 Hz and up. These two filters in series, then, will pass only frequencies between 1000 and 2000 Hz. These frequencies mark the lower and upper limits of the bandpass. This is shown in the frequency response curve of the simple bandpass filter in Fig. 4-7(c).

Bandstop LC Resonant Filter

The use of resonant LC sections offers greater Qs than the previously discussed filters. From your study of resonant circuits you know that a series resonant circuit offers minimum opposition to the resonant frequency. A circuit such as that shown in Fig. 4-8(a) would effectively short out the frequency to which the series circuit was resonant. All other frequencies would be passed from input to output. A parallel resonant circuit offers maximum opposition to the frequency of resonance. As shown in Fig. 4-8(b), the parallel resonant circuit in the line offers high opposition to the frequency to which it is resonant, and low impedance to other frequencies.

A filter made up of both series and parallel resonant circuits, such as that shown in Fig. 4-8(c), would prove effective in eliminating one side band. Let us assume that both side bands are applied to the input of this bandstop filter, one of 795 kHz and the other of 805 kHz. The balanced modulator has already eliminated the carrier frequency of 800 kHz. We wish to eliminate the upper side band (805 kHz). L_1C_1 and L_3C_3 form two series resonant circuits whose resonant frequency is 805 kHz. L_2C_2 is a parallel resonant circuit whose frequency of resonance is also 805 kHz. When the two side-band frequencies are applied to this filter, L_1C_1 effectively bypasses most of the 805-kHz signal to ground but allows the 795-kHz signal to pass. The parallel resonant circuit of L_2C_2 in the line offers maximum opposition to 805 kHz, effectively blocking it but not the 795-kHz signal. Of course, it is extremely difficult to construct a resonant circuit with a sharp enough cutoff frequency characteristic so that a small amount of the 805-kHz signal will not appear in the output. However, most of the 805-kHz signal passing the parallel resonant circuit will be eliminated by the series resonant combination of L_3C_3, which offers minimum impedance to that frequency. Now, only the lower side band, 795 kHz, will be developed across this filter circuit; the upper side band has been attenuated to the point where it is no longer of any consequence.

The Q of discrete LC circuits is limited to a maximum of several hundred, which suffices for some applications. However, most SSB systems utilize crystal,

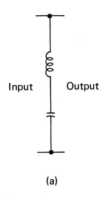

Input   Output

(a)

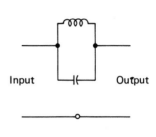

Input   Output

(b)

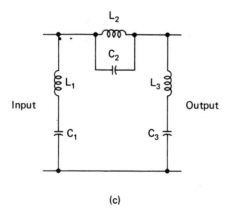

Input   Output

(c)

Fig. 4-8. Bandstop filter.

ceramic, or mechanical filters because of the much higher Q possible and better frequency stability characteristics.

Crystal Filters

The crystal filter may also be used in single-side-band systems to attenuate the unwanted side band. This is possible since the crystal is a series resonant circuit. Because of its very high Q, the crystal filter passes a much narrower band of frequencies than the best LC filter. Because of its high selectivity, it reduces the number of undesired frequencies in the output.

The equivalent circuit of the crystal and crystal holder is illustrated in Fig. 4-9(a). The components L, C_1, and R represent the series resonant circuit of the crystal itself. C_2 represents the parallel capacitance of the crystal holder. The crystal offers a very low impedance path to the frequency to which it is resonant and a high impedance path to other frequencies. However, the crystal holder capacitance, C_2, shunts the crystal and offers a path to other frequencies. For the crystal to operate as a bandpass filter, some means must be provided to counteract the shunting effect of the crystal holder. This is accomplished by placing an external variable capacitor in the circuit.

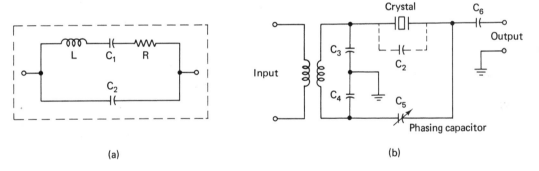

(a)   (b)

Fig. 4-9. Crystal filter.

In Fig. 4-9(b), a simple bandpass crystal filter is shown. The variable capacitor C_5, called the *phasing capacitor*, counteracts holder capacitance C_2. C_5 can be adjusted so that its capacitance equals the capacitance of C_2. Then both C_2 and C_5 pass undesired frequencies equally well. Because of the circuit arrangement, the voltages across C_2 and C_5 due to undesired frequencies are equal and 180° out of phase. Therefore, undesirable frequencies cancel, and do not appear in the output. This cancellation effect is called the *rejection notch*.

For circuit operation, assume that a lower side band with a maximum frequency of 99.9 kHz and an upper side band with a minimum frequency of 100.1 kHz are applied to the input of the crystal filter in Fig. 4-9(b). Assume that the upper side band is the unwanted side band. By selecting a crystal that will provide a low-impedance path (series resonance) at about 99.9 kHz, the lower-side-band frequency will appear in the output. The upper side band, as well as all other frequencies, will have been attenuated by the crystal filter. Crystals with a Q up to about 50,000 are available. Improved performance is possible when two or more crystals are combined in a single filter circuit.

Piezoelectric Effect

Crystals make use of the *piezoelectric effect*. This occurs in certain crystalline substances and results when a specific frequency electrical signal (the crystal's resonant frequency) is applied across the crystal and results in a mechanical oscillation of the crystal. This occurs at a specific frequency only dependent on the crystal's size and the way it is cut. In such materials, the converse effect is also observed—that a mechanical strain generates an electrical signal. Many crystalline substances exhibit this effect, but quartz crystals offer the best performance. These crystals offer far superior frequency stability compared to standard LC circuits and, as mentioned, much higher Qs. When the ultimate in frequency stability is required, they are housed in a small constant-temperature oven, since the crystal's resonant frequency varies slightly with temperature. They may have either positive or negative temperature coefficients, depending on the type of crystal cut. Their stability explains their wide use when highly accurate oscillators are required.

Ceramic Filters

Ceramic filters utilize the piezoelectric effect just as do crystals. However, they are normally constructed from lead zirconate–titanate. This is similar to the material used in ceramic phonograph cartridges.

While ceramic filters do not offer Qs as high as a crystal, they do outperform LC filters in that regard. A Q of up to 2000 is practical with ceramic filters. They are lower in cost, more rugged, and smaller in size than crystal filters. They are used not only as side-band filters but as replacements for the tuned IF transformers for superheterodyne receivers.

The circuit symbol for a ceramic filter is shown in Fig. 4-10(a) and its equivalency to the tuned transformer is indicated at B of the figure.

Mechanical Filters

Mechanical filters have been used in single-side-band equipment since the 1950s. Some of the advantages of mechanical filters are their excellent rejection characteristics, extreme ruggedness, size small enough to be compatible with the miniaturization of equipment, and a Q in the order of 10,000, which is about 50 times that obtainable with LC filters.

The mechanical filter is a device that is mechanically resonant; it receives electrical energy, converts it to mechanical vibration, then converts this mechanical energy back into electrical energy at the output. Figure 4-11 shows a cutaway view of a typical unit. There are four elements constituting a mechanical filter: (1) an

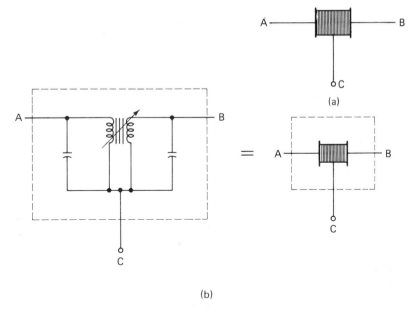

(a)

(b)

Fig. 4-10. Ceramic filter and its equivalency.

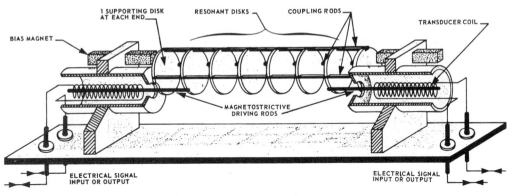

Fig. 4-11. Mechanical filter.

input transducer that converts the electrical energy at the input into mechanical vibrations, (2) metal disks that are manufactured to be mechanically resonant at the desired frequency, (3) rods that couple the metal disks, and (4) an output transducer that converts the mechanical vibrations back into electrical energy.

Not all the disks are shown in the illustration. The shields around the transducer coils have been cut away to show the coil and magnetostrictive driving rods. As you can see, either end of the filter may be used as the input.

Figure 4-12 is the electrical equivalent of the mechanical filter. The disks of the mechanical filter are represented by the series resonant circuits L_1C_1 while C_2

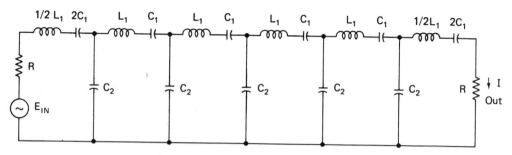

Fig. 4-12. Electrical analogy of a mechanical filter.

represents the coupling rods. The resistance R in both the input and output represents the matching mechanical loads.

Let us assume that the mechanical filter of Fig. 4-11 has disks tuned to pass the frequencies of the desired side band. The input to the filter contains both side bands, and the transducer driving rod applies both side bands to the first disk. The vibration of the disk will be greater at a frequency to which it is tuned (resonant frequency), which is the desired side band, than at the undesired side-band frequency. The mechanical vibration of the first disk is transferred to the second disk, but a smaller percentage of the unwanted side-band frequency is transferred. Each time the vibrations are transferred from one disk to the next, there is a smaller amount of the unwanted side band. At the end of the filter there is practically none of the undesired side band left. The desired side-band frequencies are taken off the transducer coil at the output end of the filter.

Varying the size of C_2 in the electrical equivalent circuit in Fig. 4-12 varies the bandwidth of the filter. Similarly, by varying the mechanical coupling between the disks (Fig. 4-11), that is, by making the coupling rods either larger or smaller, the bandwidth of the mechanical filter is varied. Because the bandwidth varies approximately as the total area of the coupling rods, the bandwidth of the mechanical filter can be increased by using either larger coupling rods or more coupling rods. Mechanical filters with bandwidths as narrow as 500 Hz and as wide as 35 kHz are practical in the 100- to 500-kHz range.

4-4 SSB TRANSMITTERS: FILTER METHOD

Figure 4-13 is a block diagram of a single-side-band transmitter using a balanced modulator to generate DSB and the filter method of eliminating one of the side bands.

For illustrative purposes, a single-tone 2000-Hz intelligence signal is used, but in actual use, a complex intelligence signal, such as the human voice produces, is typical.

The 2-kHz signal is amplified and mixed with a 100-kHz carrier or conversion frequency in the balanced modulator. Remember, neither the carrier nor audio

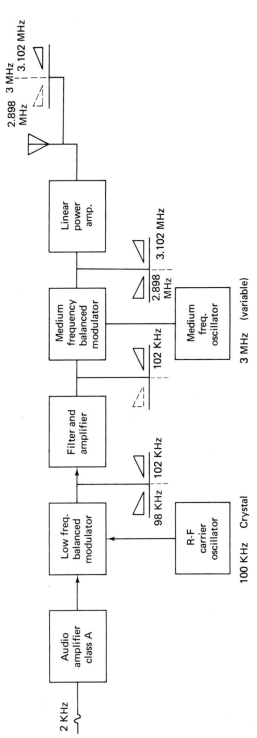

Fig. 4-13. SSB transmitter block diagram.

frequencies appear in the output of the balanced modulator; but the sum and difference frequencies (98 kHz and 102 kHz) do appear in the output. As illustrated in Fig. 4-13, the two side bands from the balanced modulator are applied to the filter. Only the desired upper side band is passed. The dotted lines show that the carrier and lower side band have been removed.

Because the remaining side band (containing the intelligence) is too low in frequency to transmit efficiently, it must be mixed again with a new conversion frequency to raise it to the desired transmitter frequency. As mentioned previously, side-band elimination at an initial low frequency allows for the use of reasonably low Q filter circuits. It also allows for the use of mechanical filters that have an upper frequency limit of about 500 kHz. In addition, crystals generally have better frequency stability at lower frequencies. The 3-MHz oscillator applies a signal to another balanced modulator. Again, the balanced modulator, after mixing the two inputs to get two new side bands, removes the new 3-MHz carrier and applies the two new side bands (3102 kHz and 2898 kHz) to a tunable linear power amplifier.

The input and output circuits of the linear power amplifier are tuned to reject one side band and pass the other to the antenna for transmission. A standard LC filter is now adequate to remove one of the two new side bands. The new side bands are about 200 kHz apart ($= 3100$ kHz $- 2900$ kHz), so the required Q is quite low. See Ex. 4-1 for further illustration. The high-frequency oscillator is variable so that the transmitter output frequency can be varied over a range of transmitting frequencies. Since both the carrier and one side band have been eliminated, all the transmitted energy is in the single side band.

EXAMPLE 4-1

For the transmitter system shown in Fig. 4-13, determine the filter Q required in the linear power amplifier.

Solution:

The second balanced modulator created another DSB signal from the SSB signal of the preceding high-Q filter. However, the frequency translation of the second balanced modulator means that a low-quality filter can be used to once again create SSB. The new DSB signal is at about 2.9 MHz and 3.1 MHz. The required filter Q is

$$\frac{3\text{ MHz}}{3.1\text{ MHz} - 2.9\text{ MHz}} = \frac{3\text{ MHz}}{0.2\text{ MHz}} = 15$$

LIC SSB Transmitter

A similar type of SSB transmitter block diagram using the previously described SL640C LIC is illustrated in Fig. 4-14. The IC is used for the initial balanced modulator and another is used for the frequency translation balanced

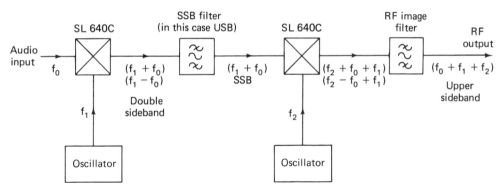

Fig. 4-14. LIC filter type SSB transmitter.

modulator. If a suppressed transmission is desired, a small amount of carrier signal is added in with the RF output shown in Fig. 4-14.

Independent Side-Band Transmitter

The independent side-band (ISB) type of SSB transmitter generates both side bands, each containing different intelligence, with the carrier suppressed to a desired level. Basically, this is accomplished with a triple conversion process that translates the two audio-frequency inputs into two RF side-band signals centered around a reinserted pilot carrier. The pilot carrier is used as a reference and to control receiving equipment auxiliary circuits such as AGC. Figure 4-15 illustrates a typical block diagram for an independent side-band transmitter.

The first conversion step converts the audio inputs (in this case, 100 Hz to 6000 Hz) into 100 kHz $\pm$ 6 kHz DSB signals. Each audio input is applied to a separate balanced modulator at the same time as the 100 kHz conversion frequency. The output of each modulator will contain the upper-side-band frequency of 100.1 to 106 kHz and the lower-side-band frequency of 94 to 99.9 kHz. The 100-kHz conversion frequency and the audio inputs do not appear in the modulator output.

Associated with each modulator is a crystal filter. The group *A* signal is fed up to the *A* crystal filter, which passes only the upper-side-band (sum frequency) output from BAL MOD 1. The group *B* signal is fed to the *B* filter, which passes only the lower-side-band (difference frequency) output from BAL MOD 2. At this point, we have succeeded in generating, or developing, the independent side bands.

This brings up another problem. How can these two signals be transmitted? To accomplish this task, a device termed a *hybrid coil* is used. This is a specially designed transformer that combines two signals without mixing or interaction. The hybrid coil is a linear device which combines the two side-band signals into one composite signal voltage that contains the two bands of frequencies of 94 to

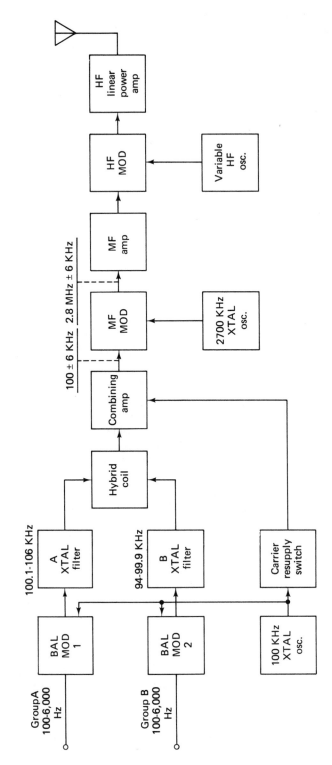

Fig. 4-15. ISB transmitter—filter type.

99.9 kHz and 100.1 to 106 kHz. The output of the hybrid coil is then applied to the combining amplifier.

The 100-kHz crystal-controlled oscillator supplies the conversion frequency for the two low-frequency modulators and the signal that will be used as the reinserted carrier. This signal is applied to the carrier resupply switch, which is adjusted for any desired level of carrier suppression. Normal independent side-band operation requires that the carrier be suppressed 20 dB with respect to the level of the side bands of intelligence. When the carrier of a transmitter rated at 4000 W peak envelope power is suppressed 20 dB, there will be approximately 40 W of power in the carrier. The remaining 3960 W of power is contained in the upper and lower side bands.

The combining amplifier in Fig. 4-15 is a class A linear amplifier. Its function is to combine the side bands with the reinserted carrier. The output of this amplifier is applied to the medium-frequency (MF) modulator, where the second frequency conversion takes place. The output of the MF modulator contains the 2600 $\pm$ 6 kHz lower side band and the 2800 $\pm$ 6 kHz upper side band. The input signal and 2700-kHz conversion frequencies are balanced out in the MF modulator circuit. The undesired modulation products are passed to ground by resonant filters, while the sum frequency is passed by a filter in the FM amplifier input. The 2800 $\pm$ 6 kHz signal is amplified by the MF amplifier and then applied to the next conversion stage.

The final frequency conversion takes place in the HF modulator when the MF signal is mixed with the conversion frequency generated by the HF oscillator, which determines the operating frequency of the transmitter. This variable oscillator makes possible an output over a wide range of frequencies in the high-frequency spectrum. Assuming that the transmitter is to operate on 5 MHz, the HF oscillator must produce an output of 7.8 MHz or 2.2 MHz, depending on whether the HF linear power amplifier is tuned to the upper- or the lower-side-band output of the HF modulator circuit. To operate on 25 MHz, the HF oscillator must produce an output of either 22.2 MHz or 27.8 MHz, depending on how the linear power amplifier is tuned. The HF linear amplifier is always tuned to the operating frequency, which will be either the upper-side-band (sum) or the lower-side-band (difference) output of the HF modulator.

4-5 SSB TRANSMITTER: THE PHASE METHOD

The phase method of SSB generation offers the following advantages over the filter method:

1. The bulk and expense of high-Q filters is eliminated.
2. Greater ease in switching from one side band to the other.
3. The ability to directly generate SSB at the desired transmitting frequency, which means that intermediate balanced modulators are not necessary.

4. Lower intelligence frequencies can be economically used with the phase method since the lower the intelligence frequency, the higher will be the Q (and thus its cost) required of the filter method.

Despite these advantages, the filter method system is rather firmly entrenched for the majority of systems, because of adequate performance for most applications and due to the complexity of the phase method.

Mathematical Analysis

The phase shift method of SSB generation relies on the fact that the upper and lower side bands of an AM signal differ in the sign of their phase angles. This means that phase discrimination may be used to cancel one side band of the DSB signal.

Consider a modulating signal $f(t)$ to be a pure cosine wave. A resulting balanced modulator output (DSB) can then be written as

$$f_{\text{DSB}1}(t) = (\cos \omega_i t)(\cos \omega_c t) \tag{4-1}$$

where $\cos \omega_i t$ is the intelligence signal and $\cos \omega_c t$ the carrier. The term $\cos A \cos B$ is equal to $\frac{1}{2}[\cos(A + B) + \cos(A - B)]$ by trigonometric identity, and therefore Eq. (4-1) can be rewritten as

$$f_{\text{DSB}1}(t) = \frac{1}{2}[\cos(\omega_c + \omega_i)t + \cos(\omega_c - \omega_i)t] \tag{4-2}$$

If another signal,

$$f_{\text{DSB}2}(t) = \frac{1}{2}[\cos(\omega_c - \omega_i)t - \cos(\omega_c + \omega_i)t] \tag{4-3}$$

were added to Eq. (4-2), the upper side band cancels, leaving just the lower side band,

$$f_{\text{DSB}1}(t) + f_{\text{DSB}2}(t) = \cos(\omega_c - \omega_i)t$$

Since the signal in Eq. (4-3) is equal to

$$\sin \omega_i t \sin \omega_c t$$

by trigonometric identity, it can be generated by shifting the phase of the carrier and intelligence signal by exactly 90° and then feeding them into a balanced modulator. Recall that sine and cosine waves are identical except for a 90° phase difference.

Block Diagram

A block diagram for the system just described is shown in Fig. 4-16. The upper balanced modulator receives the carrier and intelligence signals directly, while the lower balanced modulator receives both of them shifted in phase by 90°. Thus, combining the outputs of both balanced modulators in the adder results in a SSB output which is subsequently amplified and then driven into the transmitting antenna.

A major disadvantage of this system is involved in the 90° phase shifting network required for the intelligence signal. The *carrier* 90° phase shifter is easily

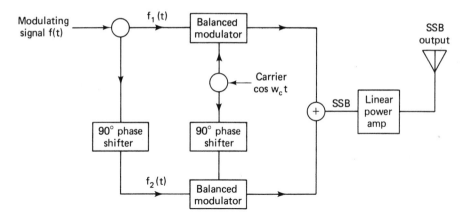

Fig. 4-16. Phase shift SSB generation.

accomplished because of its single-frequency nature, but the audio signal covers a wide range of frequencies. To obtain exactly 90° of phase shift for a complete range of frequencies is difficult. The system is critical inasmuch as an 88° phase shift (2° error) for a given audio frequency results in about 30 dB of unwanted side-band suppression instead of the desired complete suppression obtained at 90° phase shift. The difficulty in obtaining adequate performance of the intelligence phase shifting network helps explain the preference for the filter method of SSB generation.

4-6 SSB DEMODULATION

One of the major advantages of SSB has been shown to be the elimination of the transmitted carrier. We have shown that this allows an increase in effective radiated power (erp), since the side bands contain the information, and the never-changing carrier is redundant. Unfortunately, even though the carrier is redundant (contains no information), it *is* needed at the receiver! Recall that the ultimately recovered intelligence in an AM system is equal in frequency to the difference of the side-band and carrier frequencies. Thus, if a 1-kHz sine wave were to be transmitted, it creates side bands 1 kHz above and below the carrier. The received AM signal is amplified and subsequently passed through a nonlinear device (usually the diode detector) which generates sum and difference frequencies. This signal is passed through a low-pass filter to remove the high frequencies and leaves just the desired low-frequency intelligence signal.

Waveforms

Figure 4-17(a) shows three different sine-wave intelligence signals; at (b) the resulting AM waveforms are shown, and (c) shows the DSB (no carrier) waveform. Notice that the DSB envelope (drawn in for illustrative purposes) looks like a

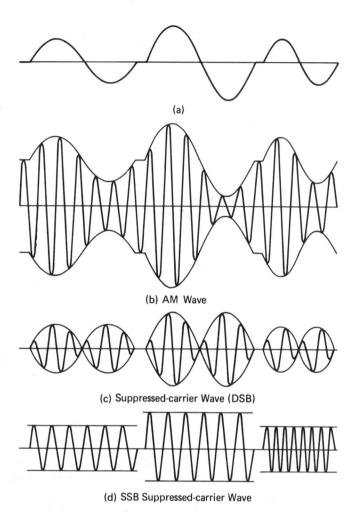

Fig. 4-17. AM, DSB and SSB wave from sinusoidal modulating signals.

full-wave rectification of the corresponding AM waveforms envelope. It is double the frequency of the AM envelope. At (d) the SSB waveforms are simply pure sine waves. This is precisely what is transmitted in the case of a sine-wave modulating signal. These waveforms are either at the carrier plus the intelligence frequency (USB) or carrier minus intelligence frequency (LSB). A SSB receiver would have to somehow "reinsert the carrier" to enable detection of the original audio or intelligence signal. A simple way to form a SSB detector is to use a mixer stage identical to a standard AM receiver mixer. The mixer is a nonlinear device, and the local oscillator input should be equivalent to the desired carrier frequency.

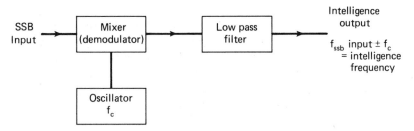

Fig. 4-18. Mixer used as SSB demodulator.

Mixer SSB Demodulator

Figure 4-18 shows this situation pictorially. Consider a 500-kHz carrier frequency that has been modulated by a 1-kHz sine wave. If the upper side band were transmitted, the receiver's demodulator would see a 501-kHz sine wave at its input. Therefore, a 500-kHz oscillator input will result in a mixer output frequency component of 1 kHz, which is the desired result. If the 500 kHz oscillator is not exactly 500 kHz, the recovered intelligence will not be exactly 1 kHz. If the receiver is to be used on several specific channels, a crystal for each channel will provide the necessary stability. If the receiver is to be used over a complete band of frequencies, the variable frequency oscillator (VFO) used must have some sort of automatic frequency control (AFC) to provide adequate quality reception. The VFO is also often termed the *beat frequency oscillator* (BFO). This can be accomplished by including a *pilot carrier* signal with the transmitted SSB signal. The pilot carrier can then be used to calibrate the receiver's oscillator at periodic intervals. Another approach is to utilize rather elaborate AFC circuits completely at the receiver, and the third possibility is the use of frequency synthesizers. They are covered in Sec. (4-8).

BFO Drift Effect

In any event, even minor drifts in oscillator frequency can cause serious problems in SSB reception. If the oscillator drifts 15 Hz, a 1-kHz intelligence signal would be detected either as 1015 Hz or 985 Hz. Speech transmission requires less than a 15-Hz shift or the talker starts sounding like Donald Duck and becomes completely unintelligible with a 40- to 50-Hz drift. Obtaining good-quality SSB reception of music and digital signals requires stabilities of several hertz.

EXAMPLE 4-2

At one instant of time, an SSB music transmission consists of a 256-Hz sine wave and its second and third harmonics, 512 Hz and 1024 Hz. If the receiver's demodulator oscillator has drifted 5 Hz, determine the resulting speaker output frequencies.

Solution:

The 5-Hz oscillator drift means that the detected audio will be 5 Hz in error, either up or down, depending on whether it is an USB or LSB transmission and on the direction of the oscillator's drift. Thus, the output would be either 251, 507, and 1019 Hz or 261, 517, and 1029 Hz. The speaker's output is no longer harmonic (exact frequency multiples), and even though it is just slightly off, the human ear would be offended by the new "music."

LIC SSB Demodulator

Other forms of SSB demodulators include modified versions of the balanced modulators used to create DSB and a simple diode detector. In any event, the frequency stability requirements of the reinjected carrier signal serve to make SSB receivers much more complex and costly than standard AM receivers.

Figure 4-19 shows the Plessey Semiconductors SL640C LIC used as a SSB detector. This is the same device previously described for the generation of DSB. The capacitor connected to output pin 5 forms the low-pass filter to allow just the audio (low)-frequency component to appear in the output.

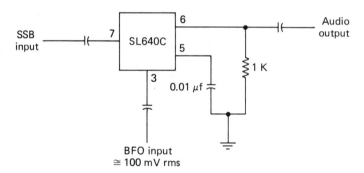

Fig. 4-19. SL640C SSB detector.

4-7 SSB RECEIVERS

To see the relationship of the parts in a single-side-band receiver, observe the block diagram in Fig. 4-20. Basically, the receiver is similar to an ordinary AM superheterodyne receiver; that is, it has RF and IF amplifiers, a mixer, detector, and audio amplifiers. However, to permit satisfactory SSB reception, an additional mixer (demodulator) and oscillator must replace the conventional diode detector.

As shown before, the carrier frequency was suppressed at the transmitter; thus, for proper intelligence detection a carrier must be inserted by the receiver. The receiver illustrated in Fig. 4-20 inserts a carrier frequency into the detector,

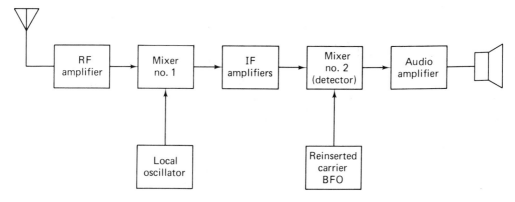

Fig. 4-20. SSB receiver block diagram.

although the carrier frequency may be inserted at any point in the receiver before demodulation of the signal.

When the SSB signal is received at the antenna, it is immediately amplified by the RF amplifier and applied to the first mixer. By mixing the output of the local oscillator with the input signal (heterodyning), a difference frequency, or IF, is obtained. The IF is then amplified by one or more stages of amplification. Of course, this is dependent upon the type of receiver. Up to this point, an AM superheterodyne receiver is exactly the same. In AM, the signal would now be applied to the detector for demodulation of the intelligence.

Second Mixer (SSB Detector)

Since the side-band frequency is the only frequency present, the carrier must now be inserted. This is accomplished by the second mixer (demodulator) and carrier generator. This latter stage is a variable-frequency oscillator that can be adjusted to simulate the carrier frequency, which was suppressed at the transmitter. Now the original intelligence can be recovered by the second mixer stage. The output of the second mixer (the intelligence) is applied to a stage of audio amplification and then to the speaker.

The second mixer is identical to the first except for the frequencies at the input and output. Consider an SSB signal of 2 MHz to be transmitted. This signal is generated by combining a carrier of 1997 kHz with an audio frequency of 3 kHz in a balanced modulator and selecting the upper side band, 1997 kHz + 3 kHz = 2 MHz (2000 kHz). When the receiver is tuned to this frequency, the 2-MHz signal is converted to a 455-kHz signal, amplified, and applied to the second mixer. When the carrier oscillator is adjusted to 458 kHz, a 3-kHz signal (458 kHz − 455 kHz) is produced in the output of the mixer.

Tuning the sideband receiver is somewhat more difficult than in a regular AM receiver. The carrier injection oscillator must be precisely adjusted to simulate

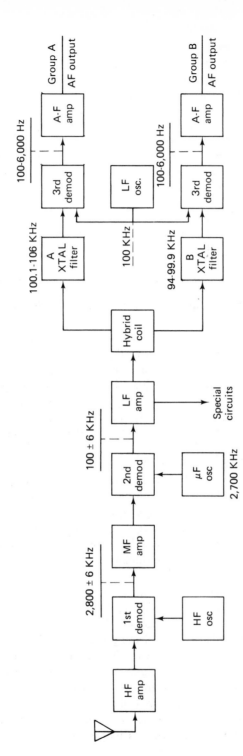

Fig. 4-21. ISB receiver block diagram.

the carrier frequency at all times. As previously explained, any tendency of drift within the oscillator will cause the output intelligence to be distorted.

ISB Receiver

In order to separate the side bands when both are transmitted, a selective crystal filter is placed in the IF amplifier strip. This filter is capable of shifting the pass band slightly above or below the normal IF frequency for reception of the upper or lower side band, respectively.

The independent side-band type of SSB receiver (Fig. 4-21) is basically the same as the SSB type except that this receiver must have two audio outputs. This is necessary to reproduce the two separate audio signals originally applied to the transmitter.

The first and second conversion stages of the independent side-band receiver are almost identical to those used by the normal communications receiver operating on the superheterodyne principle. The main difference in this receiver is the use of a hybrid coil that separates the 100 ± 6 kHz signal into two signals before application to their respective filters and demodulators. The A crystal filter passes frequencies from 100.1 to 106 kHz while the B filter passes the lower side-band frequencies from 94 to 99.9 kHz.

The signals are each demodulated by their own separate demodulator when mixed with the 100-kHz carrier frequency. The two 100- to 6000-kHz audio signals are amplified by their respective audio amplifiers and then applied to the external terminal equipment. The signal that is taken from the LF amplifier is used to operate the special circuits required for AGC and reinserted carrier applications.

Channelized SSB Equipment

SSB equipment is usually required to operate over a range of different carrier frequencies. As has already been explained, the maintenance of extremely good frequency stability between transmitter and receiver is an absolute requirement. Over the years of SSB usage, the favored method of maintaining the frequency stability has been the use of crystal oscillators, as opposed to the more troublesome pilot carrier systems. The use of crystal oscillators is ideal for single-frequency operation, but *channelized* equipment enables operation over a whole band of frequencies. *Channelized* transmitters, receivers, or *transceivers* (units that both transmit and receive) contain as many crystals as channels they are made to operate on. The appropriate crystal is selected via a switch for whatever frequency is required. This type of equipment, while still in widespread use, has the following disadvantages:

1. The one crystal for each channel requirement of necessity means the number of usable frequencies is limited.

2. Large numbers of crystals means extra bulk and cost. A typical SSB manufacturer's cost per crystal is $5.

4-8 FREQUENCY SYNTHESIS

The concept of frequency synthesis was originally begun in the 1940s. It enables the generation of literally thousands of frequencies, each with the same high accuracy of a single master oscillator. Only one crystal oscillator is required, and the extra care and cost that can then be afforded to its design leads to long-term accuracies in the area of 1 part per billion (10^9). This then means that any frequency selected from the synthesizer has an accuracy of 1 part per billion.

Synthesis is defined as formation by combining separate parts. A frequency synthesizer breaks the master reference oscillator frequency up into whole-number multiples and divisions and then combines these appropriately in mixers to provide the desired frequencies. A simple frequency synthesizer is illustrated in Fig. 4-22.

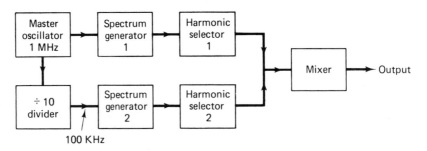

Fig. 4-22. Simple frequency synthesizer.

The master frequency feeds two separate circuits. Spectrum generator 1 takes the 1-MHz sine wave and generates a signal rich in the following harmonics: 1, 2, 3, 4, 5, 6, 7, 8, and 9 MHz. The $\div 10$ divider provides a 100-kHz output which spectrum generator 2 converts into a signal containing 100-, 200-, 300-, 400-, 500-, 600-, 700-, 800-, and 900-kHz components. The harmonic selectors have decade switches that allow only one (or none) of the nine frequency components feeding them to pass to the mixer. This selection process is determined by variable filters. The mixer then accepts the two inputs; generates components at the sum, difference, and original input frequencies; but has an output filter that accepts only the sum component.

As an example, if harmonic selector 1 is set to pass only 5 MHz and spectrum generator 2 is set to pass only 200 kHz, the output of the simple frequency synthesizer is 5 MHz + 200 kHz = 5.2 MHz.

EXAMPLE 4-3

What is the range of possible output frequencies and the "resolution" of the simple frequency synthesizer of Fig. 4-22?

Solution:

Both harmonic selectors can be set to allow no output. This would correspond to zero on their decade switches. With no input to the mixer, the output frequency is zero. Switching harmonic selector 2 to 1 allows a 100-kHz output and that same switch can be incremented to allow up to 900-kHz outputs in 100-kHz increments. Switching harmonic selector 1 to 1 and harmonic selector 2 to 0 causes an output of 1 MHz, which can be incremented to 1.9 MHz in 100-kHz steps. This process can be repeated up to a maximum 9.9 MHz of output. Thus, this frequency synthesizer has an output frequency range of from 0 Hz to 9.9 MHz with a resolution of 100-kHz increments.

Complex Synthesizer

Even the very simple frequency synthesizer just described has 100 output frequencies, with the same rock-solid stability of a single master reference oscillator. A much larger range and better resolution could be attained with the more complex synthesizer shown in Fig. 4-23.

This system allows a maximum output frequency of (900 MHz + 90 MHz + 9 MHz + 900 kHz + 90 kHz + 9 kHz) = 999.999 MHz down to 0 Hz in 1-kHz increments. It would thus allow coverage of all the popular communication bands with extremely high accuracy and frequency stability.

Frequency synthesizers are not cheap, but the cost has come down as a result of large-scale integration (LSI) usage in the relatively complex divider circuits, other IC advances for all the other circuitry, and the increasing engineering attention given to this mushrooming application. They are widely used in the electronic test equipment field, where they have replaced signal generators in those applications where greater frequency precision, which can be more quickly selected by setting a few switches, is desirable. This eliminates the time-consuming process of setting a graduated-frequency dial and checking the accuracy with a separate frequency counter. The problems of subsequent drift are also eliminated.

Synthesized Receiver

A frequency-synthesized SSB receiver is illustrated in Fig. 4-24. It is manufactured by the RF Communications Division of the Harris Corporation. The model RF-505A is a high-quality, general-purpose receiver. Performance is offered

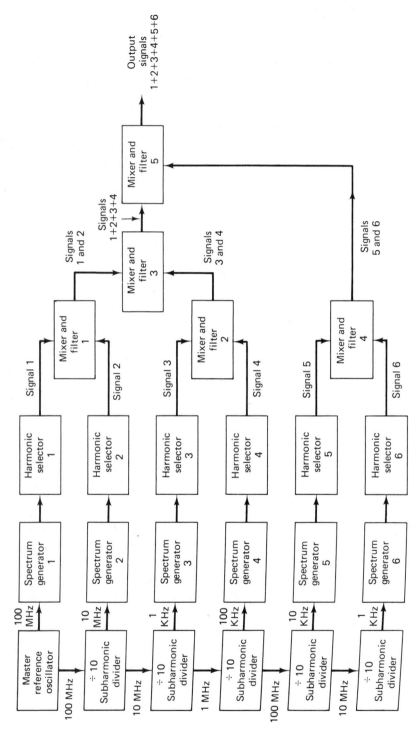

Fig. 4-23. Wide-range synthesizer.

134

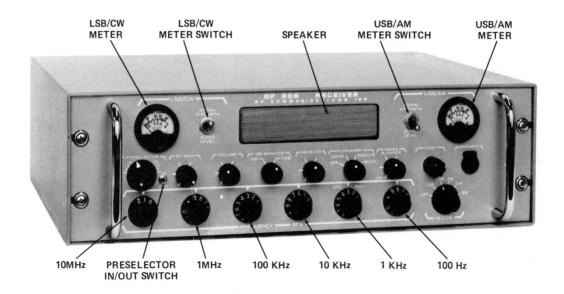

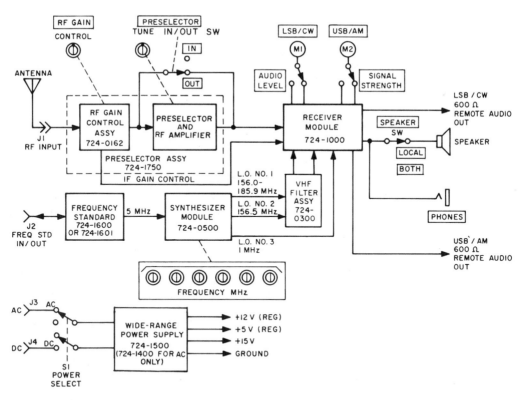

Fig. 4-24. SSB receiver. (Courtesy of Harris Corp., RF Communications Div.)

Table 1.1—RF-505A Receiver Technical Specifications*

GENERAL

Frequency Range

Up to 29.9999 MHz in synthesized 100 Hz steps. VFO continuous tuning with 10 kHz range also provided.

Frequency Stability

± 1 part in 10^6-Standard (5 MHz TCXO)

± 1 part in 10^8-Optional (with RF-508 High Stability, 5 MHz Oven)

Modes of Operation

USB, LSB, AM, CW, ISB (A1, $A3_A$, $A3_B$, $A3_G$, $A3_H$) : RATT, FAX, DATA-With external modems.

Type of Circuit

Double conversion, Superheterdyne.

RF CHARACTERISTICS

Sensitivity : 2.0000 to 29.9999 MHz

SSB/ISB : .25 uV maximum for 10 dB S+N/N in a 3 kHz bandwidth.

AM : 1.5 uV maximum, 30% modulated for 10 dB S+N/N in a 10 kHz bandwidth.

CW : .12 uV maximum for a 10 dB S+N/N in a 500 Hz bandwidth.

Note : Below 2 MHz sensitivity is gradually reduced. For example, typical SSB sensitivity at 100 kHz is 2 uV for 10 dB S+N/N.

RF Input Attenuator

0 to 20 dB, continuous.

Dynamic Range

125 dB, independent of RF gain control setting.

IF and Image Rejection

70 dB

Internal Spurious Response

Almost negligible, 99.5% below .3 uV equivalent signal at the antenna terminals.

In-Band Intermodulation

At least 45 dB below the level of two 20,000 uV emf signals in the receiver passband.

Cross Modulation

An unwanted signal of 60,000 uV, 30 kHz removed and modulated at 30% will produce cross modulation greater than 20 dB down from a desired signal of 100 uV rms, exclusive of preselection or RF gain control action. This cross modulation specification will typically remain greater than 20 dB down for all desired signals from 1 to 1,000 uV.

Blocking (Desensitization)

An undesired signal of 20,000 uV emf, 30 kHz removed shall produce less than 6 dB degradation of receiver sensitivity to the desired signal.

Input Impedance

50 ohms resistive, nominal. Less than 3 : 1 VSWR

IF CHARACTERISTICS

IF Frequencies

FIRST-156 MHz. SECOND-500 kHz.

Selectivity

SSB : Desired Sideband-300 to 3300 Hz at 6 dB ; at 5000 Hz, at least 50 dB down. Carrier-At least 25 dB down. Entire Opposite Sideband-At least 50 dB down.

AM : Nominally 6 dB at 10 kHz. At least 60 dB at 20 kHz.

CW : (with optional CW filter) 6 dB at 500 Hz. At least 60 dB at 2 kHz.

Automatic Gain Control (AGC)

Threshold : Nominally 10 uV (internally adjustable).

Range : Less than 12 dB change in output for an input signal variation of 100 dB (5 uV to 500 mV) at the input.

Attack Time : 5 milliseconds nominal.

Release Time : 1 second nominal.

AF CHARACTERISTICS

AF Response

Determined by the IF Filter in use.

AF Output

MONITOR

a) 3 watts at 5% distortion into internal speaker (3.2 ohms).

b) Speaker/Headphone monitor selection of either LSB/CW or USB/AM from the front panel.

c) Connections for remote 3.2 ohms speaker (such as the RF-511).

CHANNEL OUTPUTS

a) Two 600 ohm balanced outputs (one for each channel). Each is adjustable to 0 dBm at 5% maximum audio distortion with 10 uV input in the receiver passband.

b) Individual front panel meters are provided to monitor the audio output level or signal strength of each channel.

*Subject to change without notice.

Fig. 4-24. (Cont.).

ENVIRONMENT
Temperature
  STORAGE: OPERATING:
    −55 to +75 degrees C.   −28 to +65 degrees C.
Humidity
  95 percent
Vibration
  Per MIL-STD—167, Type 1
Shock
  Per MIL-STD-202C, Method 205C

INSTALLATION REQUIREMENTS
Wide Range Power Supply—724-1500
  AC or DC, selectable by rear panel switch.
  AC: 100 to 260 volts, single phase, 48 to 1,000
        Hz.
  DC: 10 to 40 volts, positive or negative ground.
  CONSUMPTION: 60 watts at 3 watts audio out-
                 put.

AC Power Supply 724-1400
  115/230 volts +20% −10% 48 to 65 Hz
Size
  a)   BASIC UNIT: 5 1/4H × 19 W × 13 7/8 D in.
      (13.3 H × 48.3 W × 35.2 D cm)
  b)   WITH DESK TOP CASE (RF-512): 6 H +
      19 3/8 W × 15 D in. (15.2 H × 49.2 W ×
      38.1 D cm)
  c)   WITH STACK MOUNT (RF-513): 5 1/4 H ×
      19 1/2 W × 13.7 D in. (13.3 H × 49.3 W ×
      35.2 D cm)
Weight
  a)   BASIC UNIT: 29.25 pounds (13.25 kg)
  b)   WITH DESK TOP CASE (RF-512): 33.9
      pounds (15.36 kg)
  c)   WITH STACK MOUNT (RF-513): 36 pounds
      (16.3 kg)

Fig. 4-24. (Cont.).

over the medium-and high-frequency range of 1.6 to 30 MHz, with modes of operation including USB, LSB, AM, continuous wave (CW), and ISB. Fully synthesized, the RF-505A can be digitally set to within 50 Hz of any frequency up to 29.9999 MHz. Continuous tuning is also available between each 10-kHz step.

    You may run into some unfamiliar items as you study the technical specification and block diagram shown in Fig. 4-24. These items will be studied in Chapter 7.

5

FREQUENCY MODULATION:
TRANSMISSION

5-1 ANGLE MODULATION

As has been previously stated, there are three parameters of a sine-wave carrier that can be varied to allow it to "carry" a low-frequency intelligence signal. They are its amplitude, frequency, and phase. The latter two, frequency and phase, are actually interrelated, as one cannot be changed without changing the other. They both fall under the general category of *angle modulation*. *Angle modulation* is defined as modulation in which the angle of a sine-wave carrier is the characteristic varied from its reference value. Angle modulation has two subcategories, phase modulation and frequency modulation, with the following definitions:

> *Phase modulation (PM)*: angle modulation in which the phase angle of a carrier is caused to depart from its reference value by an amount proportional to the modulating signal amplitude.
> *Frequency modulation (FM)*: angle modulation in which the instantaneous frequency of a sine-wave carrier is caused to depart from the carrier frequency

by an amount proportional to the instantaneous value of the modulating or intelligence wave.

The key difference between these two similar forms of modulation is that in PM the amount of phase change is proportional to intelligence amplitude, while in FM it is the frequency change that is proportional to intelligence amplitude. As it turns out, PM is *not* directly used as the transmitted signal in communication systems but does have importance since it is often used to help generate FM, *and* a knowledge of PM helps us to understand the superior noise characteristics of FM as compared to AM systems. In recent years, it has become fairly common practice to denote angle modulation simply as FM instead of specifically referring to FM and PM.

The concept of FM was first practically postulated as an alternative to AM in 1931. At that point, commercial AM broadcasting had been in existence for over 10 years, and the superheterodyne receivers were just beginning to supplant the TRF designs. The goal of research into an alternative to AM at that time was to develop a system less susceptible to external noise pickup. Major E. H. Armstrong developed the first working FM system in 1936, and in July 1939, he began the first regularly scheduled FM broadcast in Alpine, New Jersey.

5-2 A SIMPLE FM GENERATOR

To gain an intuitive understanding of FM, the system illustrated in Fig. 5-1 should be considered. This is actually a very simple, yet highly instructive, FM transmitting system. It consists of an *LC* tank circuit, which, in conjunction with an oscillator circuit, generates a sine-wave output. The capacitance section of the *LC* tank is not a standard capacitor but is a capacitor microphone. This popular type of microphone is often referred to as a condensor mike and is, in fact, a variable capacitor. When no sound waves reach its plates, it presents a constant value of capacitance at its two output terminals. However, when sound waves reach the mike, they

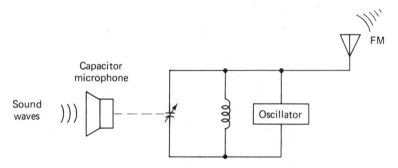

Fig. 5-1. Capacitor microphone FM generator.

alternately cause its plates to move in and out. This causes its capacitance to go up and down around its center value.

The *rate* of this capacitance change is equal to the frequency of the sound waves striking the mike *and* the *amount* of capacitance change is proportional to the amplitude of the sound waves.

Since this capacitance value has a direct effect on the oscillator's frequency, the following two *important* conclusions can be made concerning the system's output frequency:

1. The frequency of impinging sound waves determines the *rate* of frequency change.
2. The amplitude of impinging sound waves determines the *amount* of frequency change.

Consider the case of the sinusoidal sound wave (the intelligence signal) shown in Fig. 5-2(a). Up until time T_1 the oscillator's waveform at B is a constant frequency with constant amplitude. This corresponds to the carrier frequency (f_c) or *rest* frequency in FM systems. At T_1 the sound wave at A starts increasing sinusoidally and reaches a maximum positive value at T_2. During this period, the oscillator frequency is gradually increasing and reaches its highest frequency when the sound wave has maximum amplitude at time T_2. From time T_2 to T_4 the sound wave goes from maximum positive to maximum negative and the resulting oscillator frequency goes from a maximum frequency *above* the rest value to a maximum value *below* the rest frequency. At time T_3 the sound wave is passing through zero, and as such the oscillator output is instantaneously equal to the carrier frequency.

The Two Major Concepts

The amount of oscillator frequency increase and decrease around f_c is called the *frequency deviation*, δ. This deviation is shown in Fig. 5-2(c), as a function of time. It is ideally shown as a sine-wave replica of the original intelligence signal. It shows that the oscillator output is indeed an FM waveform. Recall that FM is defined as a sine-wave carrier that changes in frequency by an *amount* proportional to the instantaneous value of the intelligence wave and at a *rate* equal to the intelligence frequency.

Figure 5-2(d) shows the resulting AM wave from the intelligence signal shown at A. This should help you to see the difference between an AM and FM signal. In the case of AM, the carrier's amplitude is varied (by its side bands) in step with the intelligence, while in FM, the carrier's frequency is varied in step with the intelligence.

The capacitor microphone FM generation system is seldom used in practical applications; its importance is derived from its relative ease of providing an understanding of FM basics. If the sound-wave intelligence striking the microphone were

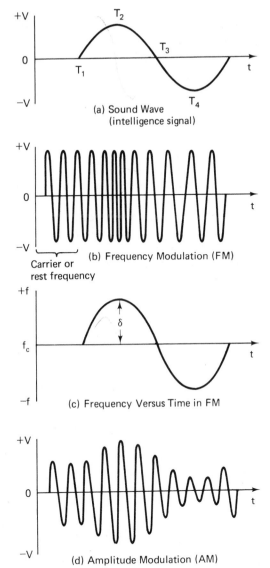

Fig. 5-2. FM representation.

doubled in frequency from 1 kHz to 2 kHz with constant amplitude, the rate at which the FM output swings above and below the center (f_c) frequency would change from 1 kHz to 2 kHz. However, since the intelligence amplitude was not changed, the *amount* of frequency deviation (δ) above and below f_c will remain the same. On the other hand, if the 1-kHz intelligence frequency were kept the same but its amplitude were doubled, the *rate* of deviation above and below f_c would remain at 1 kHz, but the *amount* of frequency deviation would double.

As you continue through your study of FM, whenever you start getting bogged down on basic theory, it will often be helpful to review the capacitor mike FM generator. *Remember*:

1. The intelligence amplitude determines the *amount* of carrier frequency deviation.
2. The intelligence frequency (f_m) determines the *rate* of carrier frequency deviation.

EXAMPLE 5-1

An FM signal has a center frequency of 100 MHz but is swinging between 100.001 MHz and 99.999 MHz at a rate of 100 times per second.

  A. Determine the intelligence frequency f_m.
  B. Determine the intelligence amplitude.
  C. Determine what happened to the intelligence amplitude if the frequency deviation changed to between 100.002 MHz and 99.998 MHz.

Solution:

  A. Since the FM signal is changing frequency at a 100-Hz rate, $f_m = 100$ Hz.
  B. There is no way of determining the actual amplitude of the intelligence signal. Every FM system has a different proportionality constant between the intelligence amplitude and the amount of deviation it causes.
  C. The frequency deviation has now been doubled, which means that the intelligence amplitude is now double whatever it originally was.

5-3 FM ANALYSIS

The complete mathematical analysis of angle modulation requires the use of high-level mathematics. For our purposes, it will suffice to simply give the solutions and discuss them. For phase modulation (PM), the equation for the instantaneous voltage is

$$e = A \sin (\omega_c t + m_p \sin \omega_m t) \tag{5-1}$$

where e = instantaneous voltage

A = peak value of original carrier wave

ω_c = carrier angular velocity ($2\pi f_c$)

m_p = maximum phase shift caused by the intelligence signal (radians)

ω_m = modulating (intelligence) signal angular velocity ($2\pi f_m$)

The maximum phase shift caused by the intelligence signal, m_p, is defined as the *modulation index* for PM.

The following equation provides the equivalent formula for FM:

$$e = A \sin (\omega_c t + m_f \sin \omega_m t) \tag{5-2}$$

All the terms in Eq. (5-2) are defined as they were for Eq. (5-1), with the exception of the new term, m_f. In fact, the two equations are identical except for that term. It is defined as the modulation index for FM, m_f. It is equal to

$$m_f = \text{FM modulation index} = \frac{\delta}{f_m} \tag{5-3}$$

where δ = maximum frequency shift caused by the intelligence signal (deviation)

f_m = frequency of the intelligence (modulating) signal

Comparison of Eqs. (5-1) and (5-2) points out the only difference between PM and FM. The equation for PM shows that the phase of the carrier varies with the modulating signal amplitude (since m_p is determined by this), and in FM the carrier phase is determined by the ratio of intelligence signal amplitude (which determines δ) to the intelligence frequency (f_m). Thus, PM is *not* sensitive to the modulating signal frequency but FM *is*. The difference between them is subtle—in fact, if the intelligence signal is integrated and then allowed to phase-modulate the carrier, an FM signal is created. This is exactly the method used in the Armstrong indirect FM system, as will be explained in Sec. (5-6).

FM Mathematical Solution

The FM formula, Eq. (5-2), is really more complex than it looks because it contains the sine of a sine. To actually solve for the frequency components of an FM wave requires the use of a high-level mathematical tool, *Bessel functions*. They show that frequency-modulating a carrier with a pure sine wave actually generates an infinite number of side bands (components) spaced at multiples of the intelligence frequency, f_m, above and below the carrier! Fortunately, the amplitude of these side bands approaches a negligible level the farther away they are from the carrier, which allows FM transmission within finite bandwidths. The Bessel function solution to the FM equation is

$$\begin{aligned}
f_c(t) = {} &J_0(m_f) \cos \omega_c t - J_1(m_f)[\cos (\omega_c - \omega_m)t - \cos (\omega_c + \omega_m)t] \\
&+ J_2(m_f)[\cos (\omega_c - 2\omega_m)t + \cos (\omega_c + 2\omega_m)t] \\
&- J_3(m_f)[\cos (\omega_c - 3\omega_m)t - \cos (\omega_c + 3\omega_m)t] \\
&+ \cdots
\end{aligned} \tag{5-4}$$

where
$$f_c(t) = \text{FM frequency components}$$

$$J_0(m_f) \cos \omega_c t = \text{carrier component}$$

$$J_1(m_f) \cos (\omega_c - \omega_m)t - \cos (\omega_c + \omega_m)t = \text{component at } \pm f_m \text{ around the carrier}$$

$$J_2(m_f) \cos(\omega_c - 2\omega_m)t + \cos(\omega_c + 2\omega_m)t = \text{component at } \pm 2f_m \text{ around the carrier, etc.}$$

In order to solve for the amplitude of any side-frequency component, J_n, the following equation should be applied:

$$J_N(m_f) = \left(\frac{m_f}{2}\right)^n \left[\frac{1}{n!} - \frac{(m_f/2)^2}{1!(n+1)!} + \frac{(m_f/2)^4}{2!(n+2)!} - \frac{(m_f/2)^6}{3!(n+1)!} + \cdots \right] \quad (5\text{-}5)$$

Thus, solving for these amplitudes is a very tedious process and strictly dependent on the modulation index, m_f. As an aid, the table shown in Table 5-1 gives the solution for a number of modulation indexes. Notice that for no modulation ($m_f = 0$), the carrier (J_0) is the only frequency present and exists at its full value of 1. However, as the carrier becomes modulated, energy is shifted from the carrier and into the side bands. For $m_f = 0.25$, the carrier amplitude has dropped to 0.98 and the first side frequencies at $\pm f_m$ around the carrier (J_1) have an amplitude of 0.12. As previously indicated, FM generates an infinite number of side bands but in this case, $J_2, J_3, J_4, \ldots$ all have negligible value. Thus, an FM transmission with $m_f = 0.25$ requires the same bandwidth ($2f_m$) as an AM broadcast.

EXAMPLE 5-2

Determine the bandwidth required to transmit an FM signal with $f_m = 10$ kHz and a maximum deviation $\delta = 20$ kHz.

Solution:

$$m_f = \frac{\delta}{f_m} = \frac{20 \text{ kHz}}{10 \text{ kHz}} = 2 \quad (5\text{-}3)$$

From Table 5-1 with $m_f = 2$, the following significant components are obtained:

$$J_0, J_1, J_2, J_3, J_4$$

This means that besides the carrier, J_1 will exist at ± 10 kHz around the carrier, J_2 at ± 20 kHz, J_3, at ± 30 kHz, and J_4 at ± 40 kHz. Therefore, the total required bandwidth is 2×40 kHz $= 80$ kHz.

EXAMPLE 5-3

Repeat Ex. 5-2 with f_m changed to 5 kHz.

Solution:

$$m_f = \frac{\delta}{f_m} \quad (5\text{-}3)$$

$$= \frac{20 \text{ kHz}}{5 \text{ kHz}}$$

$$= 4$$

TABLE 5-1

FM Side Frequencies from Bessel Functions

x (m_f)	J_0	J_1	J_2	J_3	J_4	J_5	J_6	J_7	J_8	J_9	J_{10}	J_{11}	J_{12}	J_{13}	J_{14}	J_{15}	J_{16}
0.00	1.00	—	—	—	—	—	—	—	—	—	—	—	—	—	—	—	—
0.25	0.98	0.12	—	—	—	—	—	—	—	—	—	—	—	—	—	—	—
0.5	0.94	0.24	0.03	—	—	—	—	—	—	—	—	—	—	—	—	—	—
1.0	0.77	0.44	0.11	0.02	—	—	—	—	—	—	—	—	—	—	—	—	—
1.5	0.51	0.56	0.23	0.06	0.01	—	—	—	—	—	—	—	—	—	—	—	—
2.0	0.22	0.58	0.35	0.13	0.03	—	—	—	—	—	—	—	—	—	—	—	—
2.5	−0.05	0.50	0.45	0.22	0.07	0.02	—	—	—	—	—	—	—	—	—	—	—
3.0	−0.26	0.34	0.49	0.31	0.13	0.04	0.01	—	—	—	—	—	—	—	—	—	—
4.0	−0.40	−0.07	0.36	0.43	0.28	0.13	0.05	0.02	—	—	—	—	—	—	—	—	—
5.0	−0.18	−0.33	0.05	0.36	0.39	0.26	0.13	0.05	0.02	—	—	—	—	—	—	—	—
6.0	0.15	−0.28	−0.24	0.11	0.36	0.36	0.25	0.13	0.06	0.02	—	—	—	—	—	—	—
7.0	0.30	0.00	−0.30	−0.17	0.16	0.35	0.34	0.23	0.13	0.06	0.02	—	—	—	—	—	—
8.0	0.17	0.23	−0.11	−0.29	−0.10	0.19	0.34	0.32	0.22	0.13	0.06	0.03	—	—	—	—	—
9.0	−0.09	0.24	0.14	−0.18	−0.27	−0.06	0.20	0.33	0.30	0.21	0.12	0.06	0.03	0.01	—	—	—
10.0	−0.25	0.04	0.25	0.06	−0.22	−0.23	−0.01	0.22	0.31	0.29	0.20	0.12	0.06	0.03	0.01	—	—
12.0	0.05	−0.22	−0.08	0.20	0.18	−0.07	−0.24	−0.17	0.05	0.23	0.30	0.27	0.20	0.12	0.07	0.03	0.01
15.0	−0.01	0.21	0.04	−0.19	−0.12	0.13	0.21	0.03	−0.17	−0.22	−0.09	0.10	0.24	0.28	0.25	0.18	0.12

n or Order

From E. Cambi, *Bessel Functions*, Dover Publications, Inc., New York, N.Y., 1948 (*Courtesy of the publisher*).

Referring to Table 5-1 with $m_f = 4$ shows that the highest significant side-frequency component is J_7. Since J_7 will be at $\pm 7 \times 5$ kHz
around the carrier, the required BW is 2×35 kHz = 70 kHz.

FM Bandwidth

Figure 5-3 shows the FM frequency spectrum for various levels of modulation while keeping the modulation frequency constant. The relative amplitude of
all components is obtained from Table 5-1. Notice from the table that between m_f
= 2 and $m_f = 2.5$, the carrier goes from a plus to a minus value. The minus sign
simply indicates a phase reversal, but when $m_f = 2.2$, the carrier component has
zero amplitude and all the energy is contained in the side frequencies. This also
occurs between $m_f = 5$ and 6, 7 and 9, 10 and 12, and 12 and 15.

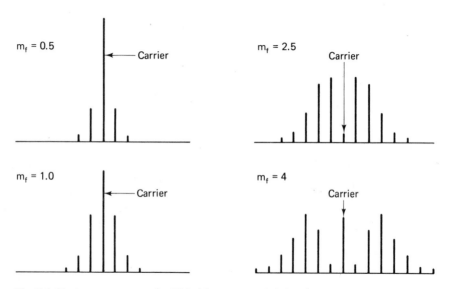

Fig. 5-3. Frequency spectrum for FM. (Constant modulating frequency,
variable deviation.)

Standard broadcast FM uses a 200-kHz bandwidth for each station. This is a
very large allocation when one considers that one FM station has a bandwidth that
could contain 20 standard AM stations with their 10-kHz bandwidths. Broadcast
FM, however, allows for a true high-fidelity modulating signal up to 15 kHz as
opposed to 5 kHz for AM and offers superior noise performance. [See Sec. (5-4).]

Figure 5-4 shows the FCC allocation for commercial FM stations. The
maximum allowed deviation around the carrier is ± 75 kHz, and 25-kHz *guard*
bands at the upper and lower ends are also provided. Recall that an infinite number
of side frequencies are generated during frequency modulation, but their amplitude
is gradually decreasing as you move away from the carrier. In other words, the

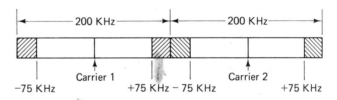

Fig. 5-4. Commercial FM bandwidth allocations for two adjacent stations.

significant side frequencies exist up to ± 75 kHz around the carrier, and the guard bands ensure that adjacent channel interference will not be a problem.

Since full deviation (δ) is 75 kHz, that is 100% modulation. By definition, 100% modulation in FM is when the deviation is the full permissible amount. Recall that the modulation index, m_f, is

$$m_f = \frac{\delta}{f_m} \qquad (5\text{-}3)$$

so that the actual modulation index at 100% modulation varies inversely with the intelligence frequency, f_m. This is in contrast with AM, where full or 100% modulation means a modulation index of 1 regardless of intelligence frequency.

Narrow-Band FM

Frequency modulation is also widely used in communication (i.e., not to entertain) systems such as police, aircraft, taxicabs, weather service, and private industry networks. These applications normally are simply voice transmissions, which means that intelligence frequency maximums of 3 kHz are the norm. These are *narrow-band* FM systems in that FCC bandwidth allocations of 10–30 kHz are provided.

EXAMPLE 5-4

    A. Determine the permissible range in maximum modulation index for commercial FM that has 30 Hz to 15 kHz modulating frequencies.

    B. Repeat for a narrow-band system that allows a maximum deviation of 10 kHz and 100 Hz to 3 kHz modulating frequencies.

Solution:

    A. The maximum deviation in broadcast FM is 75 kHz.

$$m_f = \frac{\delta}{f_m} \qquad (5\text{-}3)$$

$$= \frac{75 \text{ kHz}}{30 \text{ Hz}} = \underline{2500}$$

to
$$\frac{75\,\text{kHz}}{15\,\text{kHz}} = \underline{5}$$

B. $m_f = \dfrac{\delta}{f_m} = \dfrac{10\,\text{kHz}}{100\,\text{Hz}} = 100$

to
$$\frac{10\,\text{kHz}}{3\,\text{kHz}} = 3\tfrac{1}{3}$$

The modulation index when δ and f_m are the maximum possible value (75 kHz and 15 kHz, respectively, for broadcast FM) is often called the *deviation ratio*. It has a value of 75 kHz/15 kHz, or 5, for broadcast FM.

EXAMPLE 5-5

Determine the relative total power of the carrier and side frequencies when $m_f = 0.25$ for a 10-kW FM transmitter.

Solution:

For $m_f = 0.25$, the carrier is equal to 0.98 times its unmodulated amplitude and the only significant side band is J_1, with a relative amplitude of 0.12 (from Table 5-1). Therefore, since power is proportional to the voltage squared, the carrier power is

$$(0.98)^2 \times 10\,\text{kW} = 9.704\,\text{kW}$$

and the power of each side band is

$$(0.12)^2 \times 10\,\text{kW} = 144\,\text{W}$$

The total power is

$$9704\,\text{W} + 144\,\text{W} + 144\,\text{W} = 9.992\,\text{kW}$$
$$\cong 10\,\text{kW}$$

The result of Ex. 5-5 is predictable. In FM, the transmitted waveform never varies in amplitude, just frequency. Therefore, the total transmitted power must remain constant regardless of the level of modulation. It is thus seen that whatever energy is contained in the side frequencies has been obtained from the carrier. No additional energy is added during the modulation process. The carrier in FM is not redundant as in AM, since its (the carrier) amplitude varies dependent on the intelligence signal.

5-4 NOISE SUPPRESSION

The most important advantage of FM over AM is FM's superior noise characteristics. You are probably aware that static noise is rarely heard on FM although it is quite common in AM reception. You may be able to guess a reason for this improvement. The addition of noise to a received signal causes a change in its amplitude. Since the amplitude changes in AM contain the intelligence, any attempt to get rid of the noise adversely affects the received signal. However, in FM, the intelligence is *not* carried by amplitude changes but instead by frequency changes. The spikes of external noise picked up during transmission are "clipped" off by a *limiter* circuit and/or through the use of detector circuits that are insensitive to amplitude changes. Chapter 6 provides more detailed information on these FM receiver circuits.

Figure 5-5(a) shows the noise removal action of an FM limiter circuit, while Fig. 5-5(b) shows that the noise spike feeds right through to the speaker in an AM system. The advantage for FM is clearly evident; in fact, at first glance you may think that the limiter removes all the effects of this noise spike. Unfortunately, while it is possible to clip the noise spike off, it still causes an undesired phase shift and thus frequency shift of the FM signal, and this frequency shift *cannot* be removed.

The noise signal frequency will be close to the frequency of the desired FM signal due to the selective effect of the tuned circuits in a receiver. In other words, if you are tuned to an FM station at 96 MHz, the receiver's selectivity provides gain only for frequencies near 96 MHz. The noise that will affect this reception must, therefore, also be around 96 MHz, since all other frequencies will be greatly attenuated. The effect of adding the desired *and* noise signals will give a resultant signal with a different phase angle than the desired FM signal alone. Therefore, the noise signal, even though it is clipped off in amplitude, will cause phase modulation (PM), which indirectly causes undesired FM. The amount of frequency deviation (FM) caused by PM is

$$\delta = \phi \times f_m \tag{5-6}$$

where δ = frequency deviation

ϕ = phase shift (radians)

f_m = frequency of intelligence signal

FM Noise Analysis

The phase shift caused by the noise signal results in a frequency deviation which is predicted by Eq. (5-6). Consider the situation illustrated in Fig. 5-6. Here the noise signal is one-half the desired signal amplitude, which provides a voltage S/N ratio of 2:1. This is an untolerable situation in AM but, as the following analysis will show, is not so bad in FM. At any given instant of time, the noise and desired signal will be at a different angle with respect to each other. This is because the noise and desired signal are at different frequencies even though the receiver's tuned circuits will only allow noise signals that are within the desired station's channel allocation. The resultant of noise and desired signal is shown for a number of different angle relationships. Analysis shows that the maximum phase shift ϕ of the resultant in Fig. 5-6 occurs when the noise and desired signal are at right

150

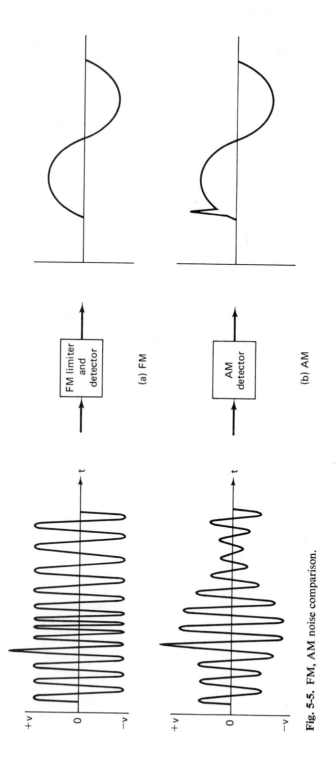

Fig. 5-5. FM, AM noise comparison.

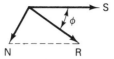
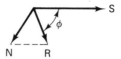

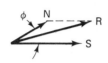
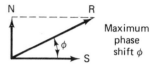

Fig. 5-6. Phase shift (ϕ) as a result of noise.

angles to each other. This is the worst-case condition. At this condition,

$$\tan \phi = \tfrac{1}{2} = 0.5$$

and therefore $\phi = 26.6°$ or about $\tfrac{1}{2}$ radian.

If the intelligence frequency, f_m, were known, then the δ caused by this severe noise condition could now be calculated using Eq. (5-6). Since $\delta = \phi \times f_m$ the worst-case deviation occurs for the maximum intelligence frequency. Assuming an f_m maximum of 15 kHz, the absolute worst case δ due to this severe noise signal is

$$\delta = \phi \times f_m = 0.5 \times 15 \text{ kHz} = 7.5 \text{ kHz}$$

In standard broadcast FM, the maximum modulating frequency is 15 kHz and the maximum allowed deviation is 75 kHz above and below the carrier. Thus, a 75-kHz deviation corresponds to maximum modulating signal amplitude and full volume at the receiver's output. The 7.5-kHz worst-case deviation output due to the S/N = 2 condition is

$$\frac{7.5 \text{ kHz}}{75 \text{ kHz}} = \frac{1}{10}$$

and, therefore, the 2:1 signal-to-noise ratio results in an output signal-to-noise ratio of 10:1. This result assumes that the receiver's internal noise is negligible. Thus, FM is seen to exhibit a very strong capability to nullify the effects of noise! In AM, a 2:1 signal-to-noise ratio at the input essentially results in the same ratio at the output. Thus, FM is seen to have an inherent noise reduction capability not possible with AM.

EXAMPLE 5-6

Determine the worst-case output S/N for a broadcast FM program that has a maximum intelligence frequency of 5 kHz. The input S/N is 2.

Solution:

The input S/N = 2 means that the worst-case deviation is $\frac{1}{2}$ radian (see the preceding paragraphs). Therefore,

$$\delta = \phi \times f_m \qquad (5\text{-}6)$$

$$= 0.5 \times 5 \text{ kHz} = 2.5 \text{ kHz}$$

Since full volume in broadcast FM corresponds to a 75-kHz deviation, this 2.5-kHz worst-case noise deviation means that the output S/N is

$$\frac{75 \text{ kHz}}{2.5 \text{ kHz}} = 30$$

Example 5-6 shows that the inherent noise reduction capability of FM is improved when the maximum intelligence (modulating) frequency is reduced. A little thought shows that this capability can also be improved by increasing the maximum allowed frequency deviation from the standard 75-kHz value. An increase in allowed deviation means that increased bandwidth for each station would be necessary, however. In fact, many FM systems utilized as communication links operate with decreased bandwidths—the narrow-band FM system. It is typical for them to operate with a 10-kHz maximum deviation. The inherent noise reduction of these systems is reduced by the lower allowed δ but is somewhat offset by the lower maximum modulating frequency of 3 kHz usually used for voice transmissions.

EXAMPLE 5-7

Determine the worst-case output S/N for a narrow-band FM receiver with $\delta_{\max} = 10 \text{ kHz}$ and a maximum intelligence frequency of 3 kHz. The S/N input is 3:1.

Solution:

The worst-case δ due to the noise occurs when the noise and desired signal are at 90°. Thus, $\tan \phi = \frac{1}{3} = 0.333$. Therefore,

$$\phi = 18.43° = 0.32 \text{ radian}$$

and

$$\delta = \phi \times f_m \qquad (5\text{-}6)$$

$$= 0.32 \times 3 \text{ kHz} = 1 \text{ kHz}$$

The S/N output will be

$$\frac{10 \text{ kHz}}{1 \text{ kHz}} = 10$$

and thus the input signal-to-noise ratio of 3 is transformed to 10 or higher at the output.

Capture Effect

This inherent ability of FM to minimize the effect of undesired signals (noise in the previous paragraphs) also applies to the reception of an undesired station operating at the same or nearly the same frequency as the desired station. This is known as the *capture effect*. You may have noticed when riding in a car that a desired FM station is suddenly replaced by a different one. You may even find that the receiver alternates abruptly back and forth between the two. This occurs because the two stations are presenting a variable signal as you drive. The capture effect causes the receiver to lock on the stronger signal by suppressing the weaker but can fluctuate back and forth when the two are of nearly equal strength. However, when they are not equal, the inherent FM noise suppression action is very effective in preventing the interference of an unwanted (but weaker) station. The weaker station is suppressed just as noise was in the preceding noise discussion. In AM, it is not uncommon to hear two separate broadcasts at the same time, but this is certainly a rare occurrence with FM.

Preemphasis

The noise suppression ability of FM has been shown to decrease with higher intelligence frequencies. This is unfortunate since the higher intelligence frequencies tend to be of lower amplitude than the low frequencies. Thus, a high-pitched violin note that the human ear may perceive as the same "sound" level as the crash of a base drum may have only half the electrical amplitude as the low-frequency drum signal. In FM, half the amplitude means only half the deviation and, subsequently, half the noise reduction capability. To counteract this effect, virtually all FM transmissions provide an artificial boost to the electrical amplitude of the higher frequencies. This process is termed preemphasis.

By definition, preemphasis is the increasing of the relative strength of the high-frequency components of the audio signal before it is fed to the modulator. Stated differently, preemphasis provides to the upper audio-frequency range the desired signal level, with respect to desired signal level in the lower-audio-frequency range. Thus, the undesirable relationship between the high-frequency intelligence components and the noise is altered. While the noise remains the same, the desired signal strength is increased. There is a disadvantage, however, in that the natural balance between high- and low-frequency tones at the receiver is altered.

A *deemphasis* circuit in the receiver, however, corrects this defect, as it reduces the high-frequency audio the same amount as the preemphasis circuit increased it, thus regaining the original tonal balance. In addition, the deemphasis network operates on both the high-frequency signal and the high-frequency noise; therefore, there is no change in the improved signal-to-noise ratio.

The main reason for the preemphasis network, then, is to prevent the high-frequency components of the transmitted intelligence from being blocked out by noise that would otherwise have more effect on the higher intelligence frequencies.

The deemphasis network is normally inserted between the detector and the audio amplifier in the receiver. This ensures that the audio frequencies are returned to their original relative level before amplification. The preemphasis characteristic curve is flat up to 500 Hz, as shown in Fig. 5-7. From 500 to 15,000 Hz there is a

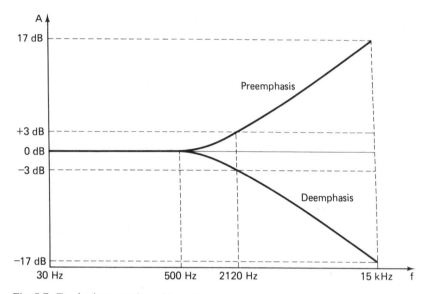

Fig. 5-7. Emphasis curves ($\tau = 75$ μsec).

sharp increase in gain up to approximately 17 dB. The gain at these frequencies is necessary to maintain the high signal-to-noise ratio at high audio frequencies. The frequency characteristic of the deemphasis network is directly opposite to that of the preemphasis network. The high-frequency response decreases in proportion to its increase in the preemphasis network. The characteristic curve of the deemphasis circuit should be a mirror image of the preemphasis characteristic curve. Figure 5-7 shows the pre- and deemphasis curves as used by standard FM broadcast in the United States. As shown, the 3-dB points occur at 2120 Hz as predicted by the *RC* time constants of 75 μs used to generate them.

$$f = \frac{1}{2\pi RC} = \frac{1}{2\pi \times 75\mu s} = 2120 \text{ Hz}$$

The FCC ruled in 1974 that FM broadcast stations can, if desired, use a 25-μs time constant. They then must use the *Dolby* noise reduction system, which works exactly like preemphasis but in a dynamic fashion. The amount of preemphasis (and subsequent deemphasis in a "dolbyized" receiver) varies depending upon the loudness level at any instant. During loud music, the noise is masked anyway and emphasis is, therefore, minimized, but the softer high-frequency passages are emphasized so that they too may "override" the noise. The change to a 25-μs time constant allows for an increased dynamic range of the Dolby effect. Maximum benefit of this system is realized by listeners in "fringe" areas where noise effects are most detrimental. Those people with regular 75-μs, non-Dolbyized receivers do not seem to be adversely affected but will usually turn down their treble control somewhat. This same Dolby noise reduction system is highly effective in minimizing high-frequency tape hiss noise in quality tape recorders.

Figure 5-8(a) shows a typical preemphasis circuit. The impedance to the audio voltage is mainly that of the parallel circuit of C and R_1, as the effect of R_2 is small in comparison to that of either C or R_1. Since

$$X_c = \frac{1}{2\pi f C}$$

audio frequency increases cause the reactance of C to decrease. This decrease of

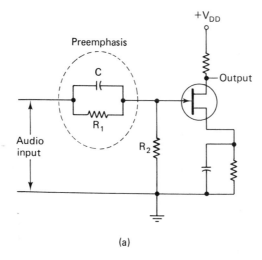

(a)

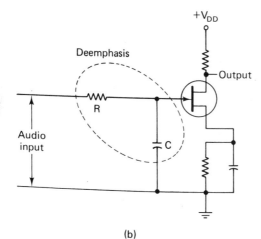

(b)

Fig. 5-8. Emphasis circuits.

X_c provides an easier path for high frequencies as compared to R. Thus, with an increase of audio frequency, there is an increase in signal voltage. The result is a larger voltage drop across R_2 (the amplifier's input) at the higher frequencies and thus greater output.

Figure 5-8(b) depicts a typical deemphasis network. Note the physical position of R and C in relation to the base of the transistor. As the frequency of the audio signal increases, the reactance of capacitor C decreases. The voltage division between R and C now provides a smaller drop across C. The audio voltage applied to the base decreases; therefore, a reverse of the preemphasis circuit is accomplished. For the signal to be exactly the same as before preemphasis and deemphasis, the time constants of the two circuits must be equal to each other.

5-5 DIRECT FM GENERATION

The capacitance microphone system explained in Sec. (5-1) can be used to generate FM directly. Recall that the capacitance of the microphone is varied in step with the sound waves striking it. Reference to Fig. 5-1 shows that if the amplitude of sound striking it is increased, the amount of deviation of the oscillator's frequency is increased. If the frequency of sound waves is increased, the rate of the oscillator's deviation is increased. This system is useful in explaining the principles of an FM signal but is not normally used to generate FM in practical systems. It is not able to produce enough deviation as required in actual applications.

Varactor Diode

A varactor diode may be used to directly generate FM. All reverse-biased diodes exhibit a junction capacitance that varies inversely with the amount of reverse bias, as was shown in Fig. 3-16. A diode that is physically constructed so as to enhance this characteristic is termed a *varactor diode*. Figure 5-9 shows a sche-

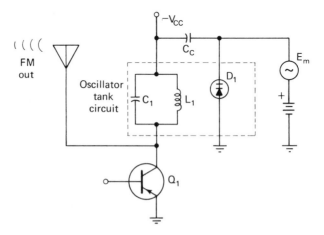

Fig. 5-9. Varactor diode modulator.

matic of a varactor diode modulator. With no modulating signal applied, the parallel combination of C_1, L_1, and D_1's capacitance forms the resonant carrier frequency. The coupling capacitor C_c isolates the dc levels and intelligence signal while looking like a short to the high-frequency carrier. When the intelligence signal, E_m, is applied to the varactor diode, its reverse bias is varied, which causes the diode's junction capacitance to vary in step with E_m. The oscillator frequency is subsequently varied as required for FM, and the FM signal is available at Q_1's collector.

Reactance Modulator

While the varactor diode modulator can be called a *reactance modulator*, the term is usually applied to those in which an active device is made to look like a variable reactance. The reactance modulator is a very popular means of FM generation. In order to determine how an active device can be made to look like a reactance, consider the JFET in Fig. 5-10. The impedance, z, looking back into

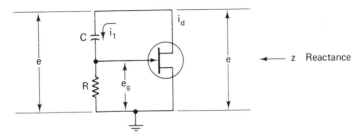

Fig. 5-10. Reactance circuit.

the JFET's drain will be shown to be reactive in this discussion. Assuming the JFET's gate current to be nearly zero,

$$e_g = i_1 R \tag{5-7}$$

but i_1 is

$$i_1 = \frac{e}{R - jX_c} \tag{5-8}$$

and substituting Eq. (5-8) into Eq. (5-7), we have

$$e_g = \frac{R \times e}{R - jX_c} \tag{5-9}$$

The JFET drain current, i_d, is

$$i_d = g_m e_g \tag{5-10}$$

(where g_m is the JFET's transconductance), which, using Eq. (5-9), can be rewritten as

$$i_d = \frac{g_m \times R \times e}{R - jX_c} \tag{5-11}$$

Therefore, the impedance, z, seen from the drain to ground will be

$$z = \frac{e}{i_d} = e \div \frac{g_m \times R \times e}{R - jX_c} \tag{5-12}$$

which can be rewritten as

$$z = \frac{R - jX_c}{g_m R} = \frac{1}{g_m} - \frac{jX_c}{g_m R} \tag{5-13}$$

If the values of R and C are chosen so that $R \ll X_c$, Eq. (5-13) reduces to

$$z = -j\frac{X_c}{g_m R} = \frac{1}{2\pi f g_m RC} \tag{5-14}$$

which can be rewritten as

$$z = \frac{1}{2\pi f C_{eq}} \tag{5-15}$$

where

$$C_{eq} = g_m RC \tag{5-16}$$

Thus, the impedance z has been made to look like a capacitance. If a modulating signal is applied to the JFET gate in Fig. 5-10, the amount of capacitance will vary in step because the JFET transconductance, g_m, is varied by an applied gate voltage. All that is necessary to generate FM, then, is to connect a JFET's drain or BJT's collector to ground terminals across an oscillator's tank circuit to provide an FM generator, as shown in Fig. 5-11. The active device can be either a tube, FET or BJT. Interchanging the position of R and C causes the variable impedance to look inductive rather than capacitive but does not stop the generation of FM.

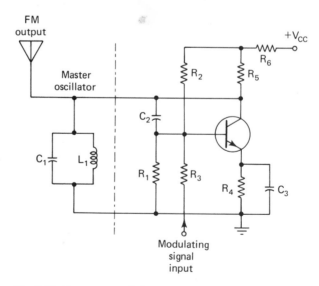

Fig. 5-11. Reactance modulator.

LIC VCO FM Generation

A *voltage-controlled-oscillator* (VCO) produces an output frequency that is directly proportional to a control voltage level. The circuitry necessary to produce such an oscillator with a high degree of linearity between control voltage and frequency was formerly prohibitive on a discrete component basis, but now that low-cost monolithic LIC VCOs are available, they make FM generation extremely simple. Figure 5-12 provides the specifications for the Signetics 566 VCO. The circuit shown at the end of the specifications (Signetics Figure 1) provides a high-quality FM generator with the modulating voltage applied to C_2. The FM output can be taken at pin 4 (triangle wave) or pin 3 (square wave). Feeding either of these two outputs into an LC tank circuit resonant at the VCO center frequency (i.e., carrier) will subsequently provide a standard sinusoidal FM signal by the flywheel effect.

Crosby Modulators

Now that three practical methods of FM generation have been shown— varactor diode, reactance modulator, and the VCO—it is time to consider the weakness of these methods. Notice that in no case was a crystal oscillator used as the basic reference or carrier frequency. The stability of the carrier frequency is very tightly controlled by the FCC, and that stability is not attained by any of the methods thus far described. Because of the high Q of crystal oscillators, it is not possible to directly frequency-modulate them—their frequency cannot be made to deviate sufficiently to provide workable wideband FM systems. It is possible to directly modulate a crystal oscillator in some narrow-band applications. If a crystal is modulated to a deviation of ± 50 Hz around a 5-MHz center frequency and both are multiplied by 100, a narrow-band system with a 500 MHz carrier ± 5 kHz deviation results. One method of circumventing this dilemma for wideband systems is to provide some means of *automatic frequency control* (AFC) to correct any carrier drift by comparing it to a reference crystal oscillator.

FM systems utilizing direct generation with AFC are called *Crosby* systems. A Crosby direct FM transmitter for a standard broadcast station at 90 MHz is shown in Fig. 5-13. Notice that the reactance modulator starts at an initial center frequency of 5 MHz and has a maximum deviation of ± 4.167 kHz. This is a typical situation in that reactance modulators cannot provide deviations exceeding about ± 5 kHz and still offer a high degree of linearity (i.e., Δf directly proportional to the modulating voltage amplitude). Consequently, *frequency multipliers* are utilized to provide a $\times 18$ multiplication up to a carrier frequency of 90 MHz (18 $\times$ 5 MHz) with a ± 75 kHz (18 $\times$ 4.167 kHz) deviation. Notice that both the carrier and deviation are multiplied by the multiplier.

Frequency multiplication is normally obtained in steps of $\times 2$ or $\times 3$ (doublers or triplers). The principle involved is to feed a frequency rich in harmonic distortion (i.e., from a class C amplifier) into an LC tank circuit tuned to two or

LINEAR INTEGRATED CIRCUITS

DESCRIPTION

The SE/NE 566 Function Generator is a voltage controlled oscillator of exceptional stability and linearity with buffered square wave and triangle wave outputs. The frequency of oscillation is determined by an external resistor and capacitor and the voltage applied to the control terminal. The oscillator can be programmed over a ten to one frequency range by proper selection of an external resistance and modulated over a ten to one range by the control voltage, with exceptional linearity.

FEATURES

- WIDE RANGE OF OPERATING VOLTAGE (10 to 24 volts)
- VERY HIGH LINEARITY OF MODULATION
- EXTREME STABILITY OF FREQUENCY (100 ppm/°C typical)
- HIGHLY LINEAR TRIANGLE WAVE OUTPUT
- HIGH ACCURACY SQUARE WAVE OUTPUT
- FREQUENCY PROGRAMMING BY MEANS OF A RESISTOR, CAPACITOR, VOLTAGE OR CURRENT
- FREQUENCY ADJUSTABLE OVER 10 TO 1 RANGE WITH SAME CAPACITOR

APPLICATIONS

TONE GENERATORS
FREQUENCY SHIFT KEYING
FM MODULATORS
CLOCK GENERATORS
SIGNAL GENERATORS
FUNCTION GENERATORS

PIN CONFIGURATION (Top View)

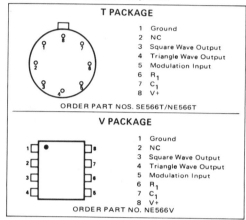

T PACKAGE

1 Ground
2 NC
3 Square Wave Output
4 Triangle Wave Output
5 Modulation Input
6 R_1
7 C_1
8 V+

ORDER PART NOS. SE566T/NE566T

V PACKAGE

1 Ground
2 NC
3 Square Wave Output
4 Triangle Wave Output
5 Modulation Input
6 R_1
7 C_1
8 V+

ORDER PART NO. NE566V

BLOCK DIAGRAM

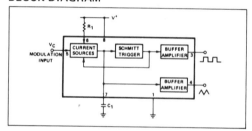

EQUIVALENT CIRCUIT

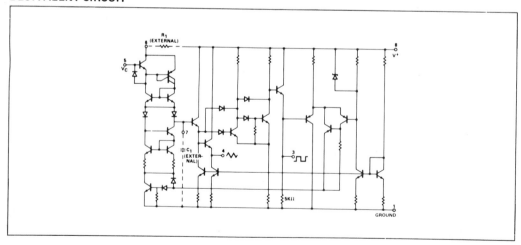

Fig. 5-12. 566 VCO specifications. (Courtesy of Signetics.)

160

SE/NE 566 — FUNCTION GENERATOR

ABSOLUTE MAXIMUM RATINGS (Limiting values above which serviceability may be impaired)

Maximum Operating Voltage	26V
Storage Temperature	−65°C to 150° C
Power Dissipation	300mW

ELECTRICAL CHARACTERISTICS (25°C, 12 Volts, unless otherwise stated)

CHARACTERISTICS	SE566			NE566			UNITS
	MIN.	TYP.	MAX.	MIN.	TYP.	MAX.	
GENERAL							
Operating Temperature Range	-55		125	0		70	°C
Operating Supply Voltage			24			24	Volts
Operating Supply Current		7	12.5		7	12.5	mA
VCO (Note 1)							
Maximum Operating Frequency		1			1		MHz
Frequency Drift with Temperature		100			200		ppm/°C
Frequency Drift with Supply Voltage		1			2		%/volt
Control Terminal Input Impedance (Note 2)		1			1		MΩ
FM Distortion (±10% Deviation)		0.2	0.75		0.2	1.5	%
Maximum Sweep Rate		1			1		MHz
Sweep Range		10:1			10:1		
OUTPUT							
Triangle Wave Output -							
Impedance		50			50		Ω
Voltage	2	2.4		2	2.4		Volts pp
Linearity		0.2			0.5		%
Square Wave Output -							
Impedance		50			50		Ω
Voltage	5	5.4		5	5.4		Volts pp
Duty Cycle	45	50	55	40	50	60	%
Rise Time		20			20		nsec
Fall Time		50			50		nsec

NOTES:

1. The external resistance for frequency adjustment (R_1) must have a value between 2KΩ and 20KΩ.

2. The bias voltage (Vc) applied to the control terminal (pin 5) should be in the range $3/4 \; V^+ \leqslant V_C \leqslant V^+$.

TYPICAL PERFORMANCE CHARACTERISTICS

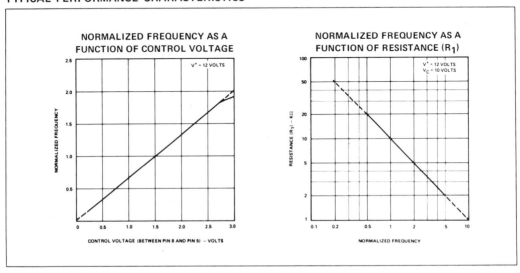

Fig. 5-12. (Cont.).

161

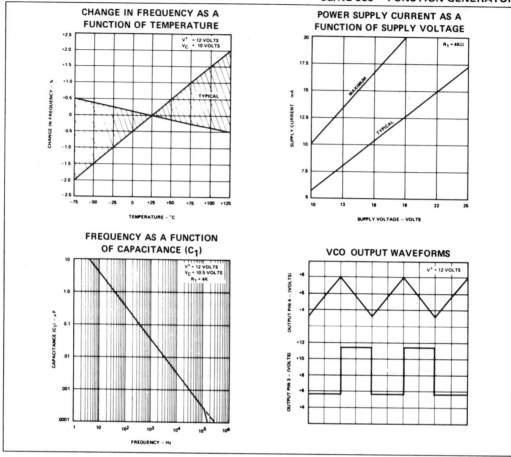

OPERATING INSTRUCTIONS

The SE/NE 566 Function Generator is a general purpose voltage controlled oscillator designed for highly linear frequency modulation. The circuit provides simultaneous square wave and triangle wave outputs at frequencies up to 1 MHz. A typical connection diagram is shown in Figure 1. The control terminal (pin 5) must be biased externally with a voltage (V_C) in the range

$$3/4 \, V^+ \leqslant V_C \leqslant V^+$$

where V_{CC} is the total supply voltage. In Figure 1, the control voltage is set by the voltage divider formed with R_2 and R_3. The modulating signal is then ac coupled with the capacitor C_2. The modulating signal can be direct coupled as well, if the appropriate dc bias voltage is applied to the control terminal. The frequency is given approximately by

$$f_o \simeq \frac{2(V^+ - V_C)}{R_1 C_1 V^+}$$

Fig. 5-12. (Cont.).

and R_1 should be in the range $2K < R_1 < 20K\Omega$.
A small capacitor (typically $0.001\mu f$) should be connected between pins 5 and 6 to eliminate possible oscillation in the control current source.

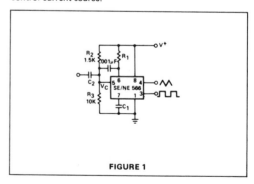

FIGURE 1

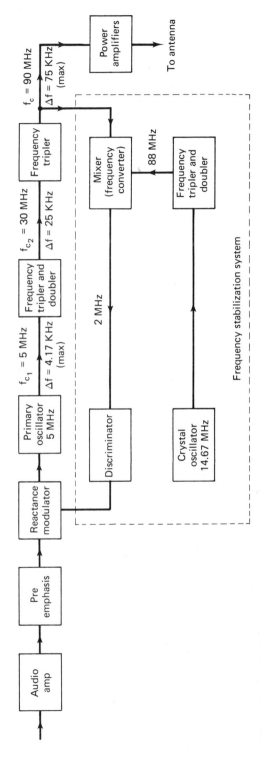

Fig. 5-13. Crosby direct FM transmitter.

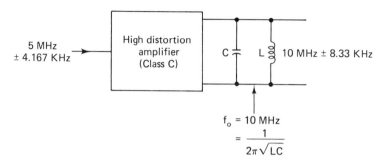

Fig. 5-14. Frequency multiplication (doubler).

three times the input frequency. The harmonic (and small deviations) are then the only significant output, as illustrated in Fig. 5-14.

After the $\times 18$ multiplication ($3 \times 2 \times 3$) shown in Fig. 5-13, the FM *exciter* function is complete. The term "exciter" is often used to denote the circuitry that generates the modulated signal. The exciter output goes to the power amplifiers for transmission *and* to the frequency stabilization system. The purpose of this system is to provide a control voltage to the reactance modulator whenever it drifts from its desired 5-MHz value. The control (AFC) voltage then varies the reactance of the primary 5-MHz oscillator slightly to bring it back on frequency.

The mixer in Fig. 5-13 has the 90-MHz carrier and 88-MHz crystal oscillator signal as inputs. The mixer output only accepts the difference component of 2 MHz, which is fed to the discriminator. A *discriminator* is the opposite of a VCO, in that it provides a dc level output based upon the frequency input. The discriminator output in Fig. 5-13 will be zero if it has an input of exactly 2 MHz, which occurs when the transmitter is at precisely 90 MHz. Any carrier drift up or down causes the discriminator output to go positive or negative, resulting in the appropriate primary oscillator readjustment. Further detail on discriminator circuits is provided in Chapter 6.

5-6 INDIRECT FM

If the phase of a crystal oscillator's output is varied, phase modulation (PM) will result. As has been previously discussed, changing the phase of a signal indirectly causes its frequency to be changed. We thus find that direct modulation of a crystal is possible via PM, which indirectly creates FM. This indirect method of FM generation is usually referred to as the *Armstrong* type, after its originator, E. H. Armstrong. It permits modulation of a stable crystal oscillator without the need for the cumbersone AFC circuitry and also provides carrier accuracies identical to the crystal accuracy, as opposed to the slight degradation of the Crosby system's accuracy.

A simple Armstrong modulator is depicted in Fig. 5-15. The JFET is biased in the ohmic region by keeping V_{DS} low. In that way it presents a resistance from drain to source that is made variable by the gate voltage that is the modulating signal. Notice that the modulating signal is first given the standard preemphasis and then applied to a frequency-correcting network. This network is a low-pass *RC* circuit (an integrator) that makes the audio output amplitude inversely proportional to its frequency. This is necessary because in phase modulation, the frequency deviation created is not only proportional to modulating signal amplitude (as desired for FM) but also to the modulating signal frequency (undesired for FM). Thus, in PM if a 1-V, 1-kHz modulating signal caused a 100-Hz deviation, a 1-V, 2-kHz signal would cause a 200-Hz deviation instead of the same deviation of 100 Hz if that signal were applied to the $1/f$ network.

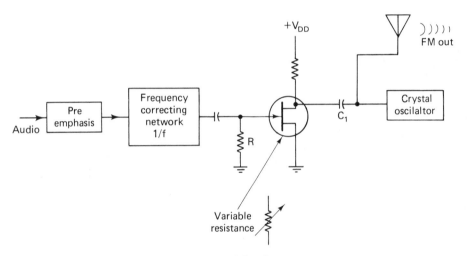

Fig. 5-15. Indirect FM via PM (Armstrong modulator).

In summary, the Armstrong modulator of Fig. 5-15 indirectly generates FM by changing the phase of a crystal oscillator's output. That phase change is accomplished by varying the phase angle of an *RC* network (C_1 and the JFET's resistance), in step with the frequency-corrected modulating signal.

Obtaining Wide-Band Deviation

The indirectly created FM is not capable of much frequency deviation. A typical deviation is 50 Hz out of 1 MHz (50 ppm). Thus, even with a $\times 90$ frequency multiplication, a 90-MHz station would have a deviation of $90 \times 50\,\text{Hz} = 4.5\,\text{kHz}$. This may be adequate for narrow-band communication FM but falls far short of the 75-kHz deviation required for broadcast FM. A complete Armstrong FM

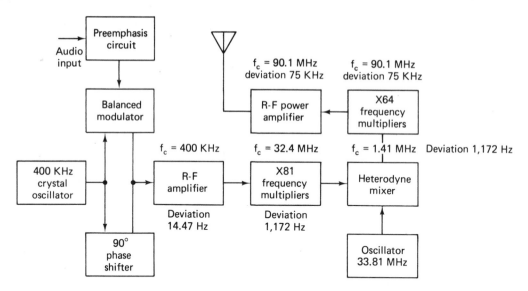

Fig. 5-16. Wideband Armstrong FM.

system providing a 75-kHz deviation is shown in Fig. 5-16. It uses the balanced modulator and 90° phase shifter to phase-modulate a crystal oscillator. Sufficient deviation is obtained by a combination of multipliers and mixing. The ×81 multipliers (3 × 3 × 3 × 3) raise the initial 400 kHz ± 14.47 Hz signal to 32.4 MHz ± 1172 Hz. The carrier *and* deviation are multiplied by 81. Applying this signal to the mixer, which also has a crystal oscillator signal input of 33.81 MHz, provides an output component (among others) of 33.81 MHz − (32.4 MHz ± 1172 Hz) or 1.41 MHz ± 1172 Hz. Notice that the mixer output changes the center frequency *without* changing the deviation. Following the mixer, the ×64 multipliers accept only the mixer difference output component of 1.41 MHz ± 1172 Hz and raise that to (64 × 1.41 MHz) ± (64 × 1172 Hz) or the desired 90 MHz ± 75 kHz.

5-7 PHASE-LOCKED LOOP FM TRANSMITTER

Block Diagram

The block diagram shown in Fig. 5-17 provides a very practical way to fabricate an FM transmitter. The amplified audio signal is used to frequency-modulate a crystal oscillator. The crystal frequency is "pulled" slightly by the variable capacitance exhibited by the varactor diode. The approximate ± 200-Hz deviation possible in this fashion is adequate for narrow-band systems. The FM output from the crystal oscillator is then divided by 2 and applied as one of the inputs to the phase detector of a phase-locked loop (PLL) system. As indicated in

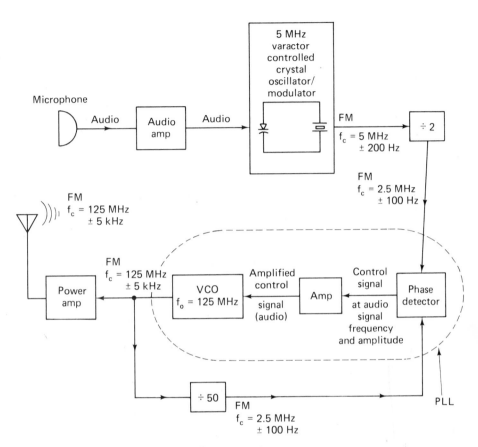

Fig. 5-17. PLL FM transmitter-block diagram.

Fig. 5-17, the other input to the phase detector is the same and its output is therefore (in this case) the original audio signal. The input control signal to the VCO is therefore the same audio signal and its output will be its free-running value of 125 MHz $\pm$ 5 kHz, which is set up to be exactly 50 times the 2.5-MHz value of the divided-by-2 crystal frequency of 5 MHz.

    The FM output signal from the VCO is given power amplification and then driven into the transmitting antenna. This output is also sampled by a $\div 50$ network, which provides the other FM signal input to the phase detector. The PLL system effectively provides the required $\times 50$ multiplication but more importantly provides the transmitter's required frequency stability. Any drift in the VCO center frequency causes an input to the phase detector (input 2 in Fig. 5-17) that is slightly different from the exact 2.5-MHz crystal reference value. The phase detector output therefore develops an error signal that corrects the VCO center frequency output back to exactly 125 MHz. This dynamic action of the phase detector/VCO and feedback path is the basis of a PLL. More detail on PLL action is provided in Chapter 6.

Circuit Description

The circuit schematic shown in Fig. 5-18 is a practical working system for the block diagram of Fig. 5-17. This circuitry approach eliminates the cumbersome oscillator–multiplier chain approach and allows for a very compact transmitter. The microphone input is amplified by U_5, which is an RCA CA3130 IC. Its output is used to "pull" the varactor-controlled crystal oscillator package comprised of Y_1, CR_3, and Q_3. Its output is amplified by Q_4 and then divided by 2 by the U_{3A} IC. Refer back to the block diagram in Fig. 5-17 as an aid in this circuit description. The output of the U_{3A} is applied as one of the inputs to the U_4 phase detector amplifier IC. The U_4 output at pin 8 is applied to the VCO made up of varactor diode CR_2 and Q_1. Its output (about 200 mW) is applied to the power amplifier stage (Q_2), which provides about 2 W into the antenna. Its output is sampled by the U_2 IC, which is an emitter-coupled logic (ECL) device that provides a $\div 10$ function at frequencies up to 250 MHz. Its output is then $\div 5$ by the TTL U_{3B} IC. Its output is then applied as the other input to the U_4 phase detector/amplifier IC. The values in this schematic are selected for operation on the 146-MHz amateur band, but operation can be accomplished at up to 250 MHz, with the U_2 ECL divider IC being the limiting upper frequency factor.

Alignment and Operation

The reference crystal frequency is determined by dividing the desired operating frequency by 25. Varactor voltage is monitored with a VTVM or oscilloscope while C_1 is varied through its range. If the loop is locked, the varactor voltage will vary with adjustment of C_1 and should be adjusted to 2.5 V. The transmitter should be terminated in a nonreactive 50-Ω load, and the RF amplifier adjusted for maximum power output. Some means of determining deviation will be necessary, and the transmitter will then be ready for use.

Operation on Other Bands

A transmitter may be constructed for use on the 200-MHz band by redesigning the oscillator and RF amplifier tuned circuits to resonate in that band. Q_1 and Q_2 will operate efficiently at frequencies up to 400 MHz. Crystal frequency is determined in the same manner as previously indicated. If separate oscillator–amplifier modules are constructed for 144 and 200 MHz, or perhaps even 50 MHz, and switched electronically with ECL gates, it is possible to operate on several bands with the same phase-locked loop components, at a considerable cost savings. It is also possible to select a low-power oscillator and an unmodulated crystal oscillator to generate the LO signal for a receiver. ECL dividers are available which allow application of this circuit at higher frequencies, but a frequency division of

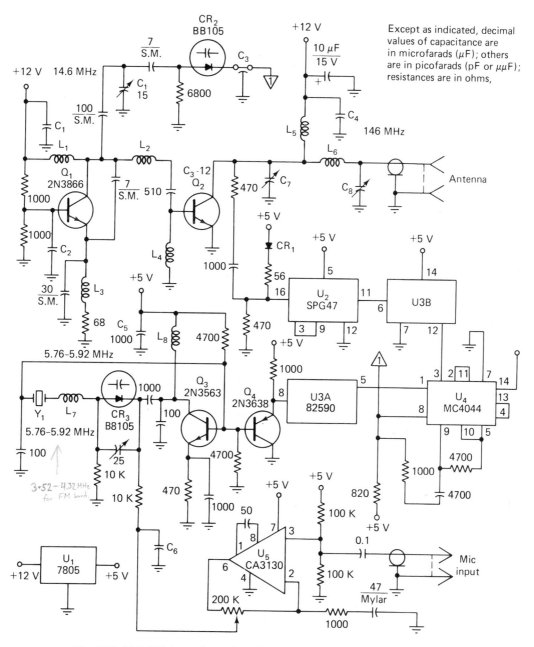

Fig. 5-18. PLL FM transmitter schematic.

C_1-C_6, incl.—1,000-pF ceramic feedthrough capacitors.

C_7, C_8—14- to 150-pF ceramic trimmer (Arco 424).

CR_2, CR_3—BB105 or Motorola MV839 Varicap diode, 82 pF nominal capacitance, 73.8- to 90.2-pF total range.

L_1, L_3, L_7—33-μH molded inductor (Miller 9230-56).

L_2—1-1/2 turns no. 20 enameled wire, 1/4-inch diameter, 1/2-inch long.

L_4—3 turns no. 28 enameled wire through ferrite bead.

L_5—2.2-μH molded inductor (Miller 9230-28).

L_6—1-1/2 turns no. 20 enameled wire, 3/8-inch diameter, 1-inch long.

L_8—100-μF molded inductor (Miller 9230-68).

Q_1—RCA 2N3866 or Motorola HEP S3008 transistor.

Q_2—C3-12, manufactured by Communications Transistor Corp., a division of Varian. An RCA 2N5913 may be substituted.

Q_3, Q_4—RCA transistor

U_1—5-volt, 1-ampere fixed positive regulator. An LM309K may be substituted.

U_2—Plessey Semiconductors integrated circuit.

U_3—Signetics 82S90 or National DM73LS196 integrated circuit.

U_4—Motorola MC4044 integrated circuit.

U_5--RCA-CA3130 integrated circuit.

Y_1—Overtone Crystal, 5.76–5.92 MHz, International Crystal Mfg. Co. Type GP. Crystal frequency is discussed in the text.

Fig. 5-18. (Cont.).

more than 50 is required in order that the maximum operating frequency of U_4 is not exceeded.

5-8 STEREO FM

The advent of stereo records and tapes and the associated high-fidelity playback equipment in the 1950s led to the development of stereo FM transmissions as authorized by the FCC in 1961. Stereo systems involve generating two separate signals, as from the left and right sides of a concert hall performance. When played back on left and right speakers, the listener gains greater spatial dimension or directivity.

A stereo radio broadcast requires that two separate 30-Hz to 15-kHz signals be used to modulate the carrier in such a way that the receiver can extract the "left" and "right" channel information and separately amplify them into their respective speakers. In essence, then, the amount of information to be transmitted is doubled in a stereo broadcast. Hartley's law (Chapter 1) tells us that either the bandwidth or time of transmission must therefore be doubled, but this is not practical. The problem was solved by making more efficient use of the available bandwidth (200 kHz) by *frequency multiplexing* the two required modulating signals. *Multiplex operation* is the simulataneous transmission of two or more signals on one carrier.

Modulating Signal

The system approved by the FCC is *compatible* in that a stereo broadcast received by a normal FM receiver will provide an output equal to the sum of the left plus right channel (L + R), while a stereo receiver can provide separate left and right channel signals. The stereo transmitter has a modulating signal, as shown in Fig. 5-19. Notice that the sum of L + R modulating signal extends from 30 Hz to 15 kHz just as does the full audio signal used to modulate the carrier in standard FM broadcasts. However, a signal corresponding to the left channel minus right channel (L − R) extends from 23 kHz to 53 kHz. In addition, a 19-kHz pilot subcarrier is included in the composite stereo modulating signal.

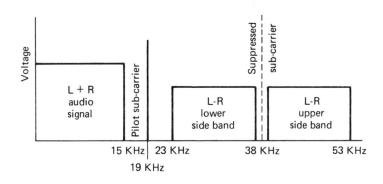

Fig. 5-19. Composite modulating signals.

The reasons for the peculiar arrangement of the stereo modulating signal will become more apparent when the receiver portion of stereo FM is discussed in Chapter 6. For now, suffice it to say that two different signals (L + R and L − R) are used to modulate the carrier. The signal is an example of *frequency-division multiplexing*, in that two different signals are multiplexed together by having them exist in two different frequency ranges.

FM Stereo Generation

The block diagram in Fig. 5-20 shows the method whereby the composite modulating signal is generated and applied to the FM modulator for subsequent transmission. The left and right channels are picked up by their respective microphones and individually preemphasized. They are then applied to a *matrix* network that inverts the right channel, giving a −R signal, and then combines (adds) L and R to provide an (L + R) signal and also combines L and −R to provide the (L − R) signal. The two outputs *are still* 30 Hz to 15 kHz audio signals at this point. The (L − R) signal and a 38-kHz carrier signal are then applied to a balanced modulator that suppresses the carrier but provides a double-side-band (DSB) signal at its output. The upper and lower side bands extend from 30 Hz to 15 kHz above and below the suppressed 38 kHz carrier and therefore range from 23 kHz (38 kHz − 15 kHz) up to 53 kHz (38 kHz + 15 kHz). Thus, the L − R signal has been translated from audio up to a higher frequency so as to keep it separate from the 30 Hz to 15 kHz (L + R) signal. The (L + R) signal is given a slight

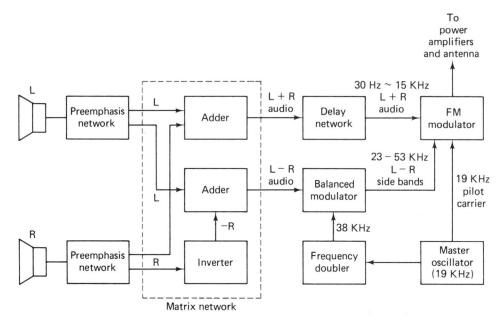

Fig. 5-20. Stereo FM transmitter.

delay so that both signals are applied to the FM modulator in time phase due to the slight delay encountered by the (L — R) signal in the balanced modulator.

The 19-kHz master oscillator in Fig. 5-20 is applied directly to the FM modulator and also doubled in frequency to 38 kHz for the balanced modulator carrier input.

The reader may wish to refer to Sec. (6-6) to directly continue this stereo FM discussion into the receiver section at this time.

5-9 FM TRANSMISSIONS

There are four major fields in which FM is widely used:

1. Commercial broadcast with 200 kHz channel bandwidths from 88 to 108 MHz.
2. Television audio signals with 50-kHz channel bandwidths at 54 to 88 MHz, 174 to 216 MHz, and 470 to 890 MHz.
3. Narrow-band public service channels from 108 MHz to 174 MHz and in excess of 890 MHz.
4. Narrow-band amateur radio channels at 29.6 MHz, 52 to 53 MHz, 146 to 147.5 MHz, 440 to 450 MHz, and in excess of 890 MHz.

The output powers range from milliwatt levels for the amateurs up to 50 kW for broadcast FM. You will note that FM is not used at frequencies below about 30 MHz. This is due to the phase distortion introduced to FM signals by the earth's ionosphere at frequencies below approximately 30 MHz. Frequencies above 30 MHz are transmitted "line-of-sight" and are not affected by the ionosphere. This situation explains the limited range (normally 70 to 80 miles) of FM transmission due to the earth's curvature.

Another advantage that FM has over SSB and AM, other than superior noise performance, is the fact that *low-level modulation* [see Sec. (2-4)] can be used with subsequent highly efficient class C power amplifiers. Since the FM waveform does not vary in amplitude, the intelligence is not lost by class C power amplification as it is for AM and SSB. Recall that a class C amplifier tends to provide a constant output amplitude due to the *LC* tank circuit flywheel effect. Thus, there is no need for high-power audio amplifiers in an FM transmitter and, more important, all the power amplification takes place at about 90% efficiency (class C) as compared to a maximum of about 70% for linear power amplifiers.

6

FREQUENCY MODULATION:

RECEPTION

6-1 BLOCK DIAGRAM

The basic FM receiver uses the superheterodyne principle. In block diagram form, it has many similarities to the receivers covered in previous chapters. In Fig. 6-1, the only apparent differences are the use of the word "discriminator" in place of "detector," the addition of a deemphasis network, and the fact that AGC may or may not be used as indicated by the dotted lines.

The *discriminator* extracts the intelligence from the high-frequency carrier and can also be called the detector, as in AM receivers. By definition, however, a discriminator is a device in which amplitude variations are derived in response to frequency or phase variations and it is the preferred term for describing an FM demodulator.

The deemphasis network following detection is required to bring the high-frequency intelligence back to the proper amplitude relationship with the lower frequencies. Recall that the high frequencies were preemphasized at the transmitter to provide them with greater noise immunity, as explained in Sec. (5-4).

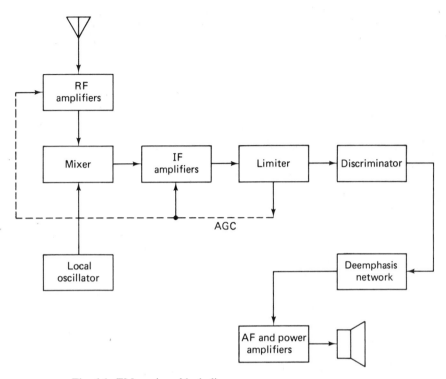

Fig. 6-1. FM receiver, block diagram.

The fact that AGC is optional in an FM receiver may be surprising to you. From your understanding of AM receivers, you know that AGC is essential to their satisfactory operation. However, the use of limiters in FM receivers essentially provides an AGC function, as will be explained in Sec. (6-3). Many older FM receivers also included an *automatic frequency control* (AFC) function. This is a circuit that provides a slight automatic control over the local oscillator circuit. It compensated for drift in LO frequency that would otherwise cause a station to become detuned. It was necessary because it had not yet been figured out how to make an economical *LC* oscillator at 100 MHz with sufficient frequency stability. If you find a new FM receiver with AFC, you can assume it to be an advertising gimmick aimed at duping the public.

The mixer, local oscillator, and IF amplifiers are basically similar to those discussed for AM receivers and do not require further elaboration. It should be noted that higher frequencies are usually involved, however, because of the fact that FM systems generally function at higher frequencies. The universally standard IF frequency for FM is 10.7 MHz, as opposed to 455 kHz for AM. Because of significant differences in all the other sections of the block diagram shown in Fig. 6-1, they are discussed in the following sections.

Broadcast AM receivers normally operate quite satisfactorily without any RF amplifier. This is rarely the case with FM receivers, however, except for frequencies in excess of 1000 MHz (1 GHz) when it becomes preferable to omit it. The essence of the problem boils down to the fact that FM receivers can function with smaller received signals than AM or SSB receivers because of their inherent noise reduction capability. This means that FM receivers can function with a lower sensitivity, and thus they are called upon to deal with input signals of 1 μV or less as compared with perhaps a 30 μV minimum input for AM. If a 1-μV signal is fed directly into a mixer, the inherently high noise factor of an active mixer stage destroys the intelligibility of the 1-μV signal. It is, therefore, necessary to amplify the 1-μV level in an RF stage to get the signal up to at least 10 to 20 μV before mixing occurs. The FM system can tolerate 1 μV of noise from a mixer on a 20-μV signal but obviously cannot cope with 1 μV of noise with a 1-μV signal.

This reasoning also explains the abandonment of RF stages for the ever-increasing FM systems at the 1-GHz-and-above region. At these frequencies, transistor noise is increasing while their gain is decreasing. The frequency is therefore reached where it is more advantageous to feed the incoming FM signal directly into a diode mixer so as to immediately step it down to a lower frequency for subsequent amplification. Diode (passive) mixers are much less noisy than active mixers.

Of course, the use of an RF amplifier reduces the image frequency problem as explained in Chapter 3. Another benefit is the reduction in *local oscillator reradiation* effects. Without an RF amp, the local oscillator signal can more easily get coupled back into the receiving antenna and transmit interference.

FET RF Amplifiers

Virtually all RF amps used in quality FM receivers utilize FETs as the active element. You may think that this is done because of their high input impedance, but this is *not* the reason. In fact, their input impedance at the high frequency of FM signals is greatly reduced because of their input capacitance. The fact that FETs do not offer any significant advantage over other devices at high frequencies is not a deterrent, however, since the impedance that an RF stage works from (the antenna) is only several hundred ohms or less anyway.

The major advantage is that FETs have a distortion input/output square-law relationship while vacuum tubes have a $\frac{3}{2}$ power relationship and BJTs have a diode-type exponential characteristic. A square-law device has an output signal at the input frequency and a smaller distortion component at two times the input frequency, whereas the other devices mentioned have many more distortion components, with some of them occurring at frequencies close to the desired signal. The use of a FET at the critical small signal level in a receiver means that the device distortion components are easily filtered out by its tuned circuits, since the closest distortion component is two times the frequency of the desired signal. This becomes

176

an extreme factor when you tune to a weak station that has a very strong adjacent signal. If the high-level adjacent signal gets through the input tuned circuit, even though greatly attenuated, it would probably generate distortion components at the desired signal frequency by a non-square-law device, and the result is audible noise in the speaker output. This form of receiver noise is called *cross-modulation*. This is similar to *intermodulation distortion*, which is characterized by the mixing of *two* undesired signals, which results in an output component that is equal in frequency to the desired. The possibility of intermodulation distortion is also greatly minimized by the use of FET RF amplifiers.

MOSFET RF Amplifiers

A dual-gate common-source MOSFET RF amplifier is shown in Fig. 6-2. The use of a dual-gate device allows a convenient isolated input for an AGC level to control device gain. The MOSFETs also offer the advantage of increased *dynamic range* over JFETs. That is, a wider range of input signal can be tolerated by the MOSFET while still offering the desired square-law input/output relationship. A similar arrangement is often utilized in mixers since the extra gate allows for a convenient injection point for the local oscillator signal. The accompanying chart in Fig. 6-2 provides component values for operation at 100 MHz and 400-MHz center frequencies. The antenna input signal is coupled into gate 1 via the coupling/tuning network comprised of C_1, L_1, and C_2. The output signal is taken at

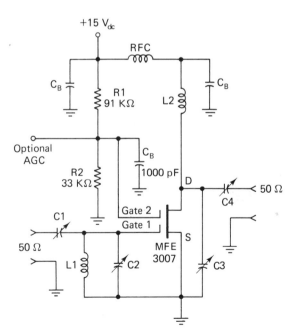

VHF Amplifier

The following component values are used for the different frequencies:

Component Values	100 MHz	400 MHz
C1	8.4 pF	4.5 pF
C2	2.5 pF	1.5 pF
C3	1.9 pF	2.8 pF
C4	4.2 pF	1.2 pF
L1	150 nH	16 nH
L2	280 nH	22 nH
C_B	1000 pF	250 pF

Fig. 6-2. MOSFET RF amplifier. (Courtesy of Motorola Semiconductor Products, Inc.)

the drain, which is coupled to the next stage by the L_2, C_3, C_4 combination. The bypass capacitor C_B next to L_2 and the radio-frequency choke (RFC) ensure that the signal frequency is not applied to the dc power supply. The RFC acts as an open to the signal while appearing as a short to dc, and the bypass capacitor acts in the inverse fashion. These precautions are necessary at RF frequencies because while power supply impedance is very low at low frequencies and dc, it looks like a high impedance to RF and can cause appreciable signal power loss. Notice, also, the bypass capacitor from gate 2 to ground, which provides a short to any signal frequency level that may get to that point. It is necessary to maintain the bias stability that is set up by R_1 and R_2. The MFE 3007 MOSFET used in this circuit provides a minimum power gain of 18 dB at 200 MHz.

6-3 LIMITERS

A limiter is a circuit whose output is a constant amplitude for all inputs above a critical value. Its function in an FM receiver is to remove any residual (unwanted) amplitude modulation and the amplitude variations due to noise. Both of these variations would have an undesirable effect if carried through to the speaker. In addition, the limiting function also provides AGC action since signals from the critical minimum value up to some maximum value provide a constant input level to the discriminator. By definition, the discriminator should ideally not respond to amplitude variations anyway, since the information is contained in the amount of frequency deviation and the rate at which it deviates back and forth around its center frequency.

A transistor limiter is shown in Fig. 6-3. Notice the dropping resistor, R_c,

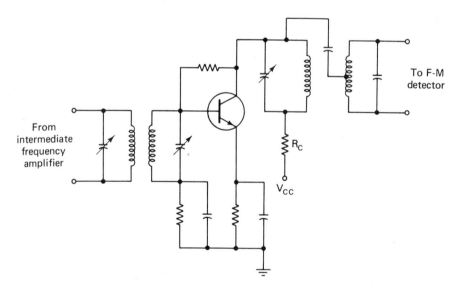

Fig. 6-3. Transistor limiting circuit.

which limits the dc collector supply voltage. This provides a low dc collector voltage which makes this stage very easily overdriven. This is the desired result. As soon as the input is large enough to cause clipping at both extremes of collector current, the critical limiting voltage has been attained and limiting action has started.

The input/output characteristic for the limiter is shown in Fig. 6-4, and it shows the desired clipping action and the effects of feeding the limited (clipped) signal into an *LC* tank circuit tuned to the signal's center frequency. The natural flywheel effect of the tank removes all frequencies not near the center frequency and thus provides a sinusoidal output signal as shown.

Limiting and Sensitivity

A limiter, such as shown in Fig. 6-3, requires about 1 V of signal to begin limiting. Much amplification of the received signal is, therefore, needed prior to limiting, which explains its position following IF amplification. When enough signal arrives at the receiver to start limiting action, the set *quiets*, which means that background noise disappears. The *sensitivity* of an FM receiver is defined in terms of how much input signal is required to produce a specific level of quieting, normally 30 dB. This means that a good-quality receiver with a rated 1.5-μV sensitivity will have background noise 30 dB down from the desired input signal that has a 1.5-μV level.

The minimum required voltage for limiting is called the *quieting, threshold,* or *limiting knee* voltage. The limiter then provides a constant-amplitude output up to some maximum value which prescribes the limiting range. Going above the maximum value results either in a reduced and/or distorted output. It is possible that a single-stage limiter will not allow for adequate range, thereby requiring a double limiter (Fig. 6-5) or the development of AGC control on the RF and IF amplifiers to minimize the possible limiter input range. The double limiter in Fig. 6-5 can provide a variable quieting voltage by base bias adjustment on either transistor.

It is most common for today's FM receivers to use IC IF amplification. In these cases, the ICs have a built-in limiting action of very high quality (i.e., wide dynamic range). Section (6-7) provides an example of these ICs.

> EXAMPLE 6-1
>
> A certain FM receiver provides a voltage gain of 200,000 (106 dB) prior to its limiter. The limiter's quieting voltage is 200 mV. Determine the receiver's sensitivity.
>
> *Solution:*
>
> In order to reach quieting, the input must be
>
> $$\frac{200 \text{ mV}}{200,000} = 1 \ \mu\text{V}$$
>
> The receiver's sensitivity is therefore 1 μV.

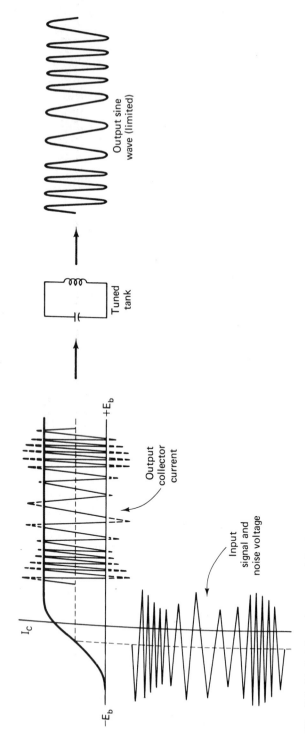

Fig. 6-4. Limiter input/output and flywheel effects.

Output sine wave (limited)

Tuned tank

Output collector current

Input signal and noise voltage

I_c

$+E_b$

$-E_b$

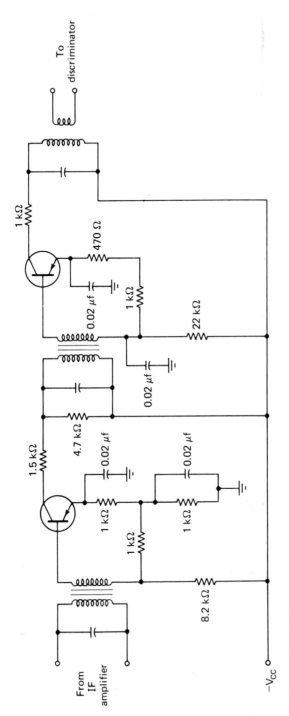

Fig. 6-5. Double limiter.

181

6-4 DISCRIMINATORS

The FM discriminator (detector) extracts the intelligence that has been modulated onto the carrier via frequency variations. It should provide an intelligence signal whose amplitude is dependent upon instantaneous carrier frequency deviation and whose frequency is dependent upon the carrier's rate of frequency deviation. A desired output amplitude versus input frequency characteristic for a broadcast FM discriminator is provided in Fig. 6-6. Notice that the response is linear in the allowed area of frequency deviation and that the output amplitude is directly proportional to carrier frequency deviation. Keep in mind, however, that FM detection takes place following the IF amplifiers, which means that the ±75-kHz deviation is intact but that carrier frequency translation (usually to 10.7 MHz) has occurred.

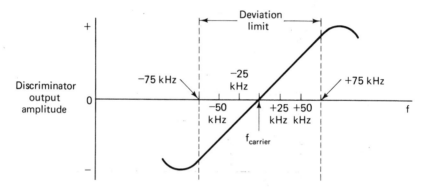

Fig. 6-6. FM discriminator characteristic.

Slope Detector

The easiest-to-visualize FM *discriminator* is the slope detector in Fig. 6-7. The *LC* tank circuit which follows the IF amplifiers and limiter is detuned from the carrier frequency so that f_c falls in the middle of the most linear region of the response curve. When the FM signal rises in frequency above f_c, the output amplitude increases while deviations below f_c cause a smaller output. The slope detector thereby changes FM into AM, and a simple diode detector then recovers the intelligence contained in the AM waveform's envelope. In an emergency, an AM receiver can be used to receive FM by detuning the tank circuit feeding the diode detector. Slope detection is not widely used in FM receivers because the slope characteristic of a tank circuit is not very linear, especially for the large-frequency deviations of wide-band FM.

Foster–Seely Discriminator

The two classical means of FM detection are the Foster–Seely discriminator and the ratio detector. While their once widespread use is now diminishing due to new techniques afforded by ICs, they remain a popular means of discrimination

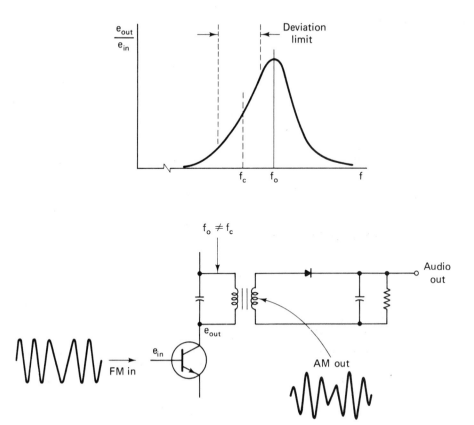

Fig. 6-7. Slope detection.

using a minimum of circuitry. A typical Foster–Seely discriminator circuit is shown in Fig. 6-8. In it, the two tank circuits $[L_1C_1$ and $(L_2 + L_3)C_2]$ are tuned exactly to the carrier frequency. Capacitors C_c, C_4, and C_5 are shorts to the carrier frequency. The following analysis applies to an unmodulated carrier input:

1. The carrier voltage e_1 appears directly across L_4 because C_c and C_4 are shorts to the carrier frequency.
2. The voltage e_s across the transformer secondary (L_2 in series with L_3) is 180° out of phase with e_1 by transformer action, as shown in Fig. 6-9(a). The circulating $L_2L_3C_2$ tank current, i_s, is in phase with e_s since the tank is resonant.
3. The current, i_s, flowing through inductance L_2L_3 produces a voltage drop that lags the current, i_s, by 90°. The individual components of this voltage, e_2 and e_3, are thus displaced by 90° from i_s, as shown in Fig. 6-9(a) and are 180° out of phase with each other because they are the voltage from the ends of a center-tapped winding.

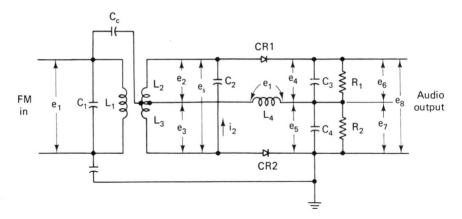

Fig. 6-8. Foster-Seely discriminator.

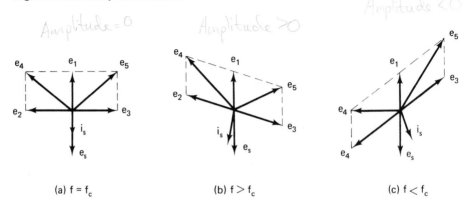

Fig. 6-9. Discriminator phase relations.

4. The voltage e_4 applied to the CR_1, C_3, and R_1 network will be the vector sum of e_1 and e_2 [Fig. 6-9(a)]. Similarly, the voltage e_5 is the sum of e_1 and e_3.

5. The output voltage, e_8, is equal to the sum of e_6 and e_7 and is zero since the diodes CR_1 and CR_2 will be conducting current equally but in opposite directions through the R_1C_3 and R_2C_4 networks.

The discriminator output is zero with no modulation (frequency deviation), as is desired. The following discussion now considers circuit action at some instant when the input signal e_1 is above the carrier frequency. The phasor diagram of Fig. 6-9(b) is used to illustrate this condition:

1. Voltages e_1 and e_s are as before, but e_s now sees an inductive reactance, because the tank circuit is above resonance. Therefore, the circulating tank current, i_s, lags e_s.

2. The voltages e_2 and e_3 must remain 90° out of phase with i_s as shown in Fig. 6-9(b). The new vector sum of e_2e_1 and e_3e_1 are no longer equal, so e_4 causes a heavier conduction of CR_1 than exists for CR_2.
3. The output, e_8, which is the sum of e_6 and e_7, will thus go positive, since the current down through R_1C_3 is greater than the current up through R_2C_4.

The output for frequencies above resonance (f_c) is therefore positive, while the phasor diagram in Fig. 6-9(c) shows that at frequencies below resonance the output goes negative. The amount of output is determined by the amount of frequency deviation, while the frequency of the output is determined by the rate at which the FM input signal varies around its carrier or center value.

Ratio Detector

While the Foster–Seely discriminator just described offers excellent linear response to wide-band FM signals, it also responds to any undesired input amplitude variations. The *ratio detector* in Fig. 6-10 does not and therefore minimizes the required limiting before detection.

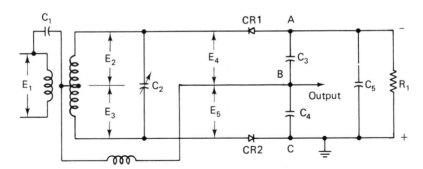

Fig. 6-10. Ratio detector.

The ratio detector, shown in Fig. 6-10, is a circuit designed to respond only to frequency changes of the input signal. Amplitude changes in the input have no effect upon the output. The input circuit of the ratio detector is identical to that of the Foster–Seely discriminator circuit. The most immediately obvious difference is the reversal of one of the diodes.

The voltages appearing across CR_1 (E_4) and CR_2 (E_5) when the input is below, at, and above the center frequency are shown in Fig. 6-11. The input circuit is the only portion of the ratio detector that operates in the same way as the Foster–Seely discriminator. CR_1, R_1, and CR_2 form a series circuit that is fed by the tuned circuit of the transformer secondary. Since the two diodes are connected in series, they conduct on the same half-cycle of the input FM signal.

The rectified current flowing through CR_1 and CR_2 produces a voltage drop across R_1 with a negative polarity at the anode of CR_1 as shown in Fig. 6-10. C_5

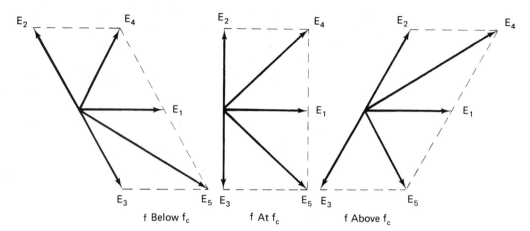

Fig. 6-11. Voltage relationships at input of ratio detector.

is a very large capacitor connected in parallel with R_1. Its function is to keep the voltage drop across R_1 constant even for the slowest-frequency deviations of the input signal. Thus, the voltage between points A and C, Fig. 6-10, remains constant. This is made possible by the large value of R_1 and C_5. This is the reason amplitude variations in the input voltage have no effect on the output.

The rectified voltage across C_3 is proportional to E_4, the voltage appearing across CR_1. The rectified voltage across C_4 is proportional to E_5, the voltage appearing across CR_2. It has been pointed out that E_4 and E_5 vary in magnitude in accordance with the frequency variations of the input signal. Thus, the voltage across C_3 and that across C_4 vary in the same manner. The sum of the voltages across the two (between points A and C) remains constant. However, the output taken at point B varies with respect to points A and C. If the potential across C_3 increases, that across C_4 decreases by the same amount, since their sum is fixed by the constant voltage across C_5. This circuit gets the name *ratio detector* because the ratio of the voltage across C_3 to that across C_4 may be continually varying, while the sum of the two voltages remains constant.

With the circuit connected as shown in Fig. 6-10, the output is always taken with respect to ground and varies about an average negative potential. The output voltage, as the frequency varies, can be determined from Fig. 6-11. Remember that the voltage across C_3 is proportional to E_5. When the input frequency is below f_c, E_5 is greater than E_4, producing a voltage across C_4 that is greater than that across C_3. At this time, then, the output is maximum negative with respect to ground, as shown in Fig. 6-12. When the input is at f_c, the voltages across C_3 and C_4 are equal since E_4 equals E_5. At this time the output is at the average negative potential. When the input frequency is above f_c, the voltage across C_3 is greater than that across C_4, since E_4 exceeds E_5. The output at this time is the least negative, as shown in Fig. 6-12.

The ratio detector provides a limiting function and is, therefore, often picked

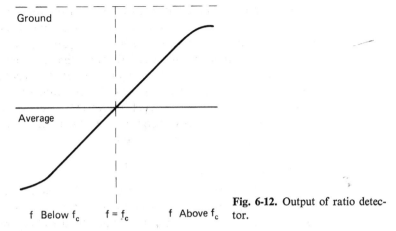

Fig. 6-12. Output of ratio detector.

over a Foster–Seely discriminator but does not offer quite as linear (distortion-free) an output or as much magnitude output. The ratio detector also can conveniently provide an AGC level from across the R_1, C_5 combination in Fig. 6-10 which is not available from a Foster–Seely discriminator.

6-5 PHASE-LOCKED LOOP

The *phase-locked loop* (PLL) has become increasingly popular as a means of FM demodulation in recent years. It eliminates the need for the intricate coil adjustments of the previously discussed discriminators and has many other uses in the field of electronics. It is an example of an old idea, originated in 1932, that has been given a new life by integrated circuit technology. Prior to its availability in a single IC package in 1970, its complexity in discrete circuitry form made it economically unfeasible for most applications.

The PLL is an electronic feedback control system as represented by the block diagram in Fig. 6-13. The input is to the *phase comparator* or *phase detector*, as it is also called.

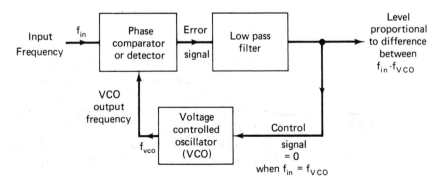

Fig. 6-13. PLL block diagram.

The VCO within the PLL forms the other input to the comparator, which looks at its two inputs and develops an output that is zero if the two input frequencies are identical. If the two input frequencies are not identical, then the comparator's output, when passed through the low-pass filter, is a level that is applied to the VCO's input. This action "closes" the feedback loop since the level applied to the VCO input changes the VCO frequency in an attempt to make it exactly match the input frequency. If the VCO frequency equals the input frequency, the PLL has achieved "lock" and the control voltage will be zero for as long as the PLL input frequency remains constant.

PLL FM Demodulator

If the PLL input is an FM signal, the low-pass filter output (VCO input) is the demodulated signal. The VCO input control signal (demodulated FM) causes the VCO output to match the FM input to the comparator. If the FM carrier (center) frequency drifts because of local oscillator drift, the PLL readjusts itself and no realignment is necessary. In a conventional FM discriminator, any shift in the FM carrier frequency results in a distorted output, since the *LC* detector circuits are then untuned. The PLL FM discriminator requires no tuned circuits and their associated adjustments, since it "adjusts" itself to any carrier frequency drifts caused by LO drift. In addition, the PLL normally has large amounts of internal amplification, which allows the input signal to vary from the microvolt region up to several volts. Since the phase comparator responds only to frequency changes and not to amplitudes, the PLL is seen to provide a limiting function of extremely wide range. The use of PLL FM detectors is widespread in current designs.

PLL Capture and Lock

Once the VCO starts to change frequency, it is in the *capture* state. It then continues to change frequency until its output is the same frequency as the input. At that point, the PLL is *locked*. The PLL has three possible states of operation:

1. Free-running.
2. Capture.
3. Locked or tracking.

If the input and VCO frequency are too far apart, the PLL free-runs at the nominal VCO frequency, which is determined by an external timing capacitor. This is not a normally used mode of operation. If the VCO and input frequency are close enough, the capture process begins and continues until the locked condition is reached. Once tracking (lock) begins, the VCO can remain locked over a wider input-frequency-range variation than was necessary to achieve capture. The tracking and capture range are a function of external resistors and/or capacitors selected by the user.

EXAMPLE 6-2

A PLL is set up such that its VCO free-runs at 10 MHz. The VCO does not change frequency until the input is within 50 kHz of 10 MHz. After that condition, the VCO follows the input to ± 200 kHz of 10 MHz before the VCO starts to free-run again. Determine the lock and capture range of the PLL.

Solution:

The capture occurred at 50 kHz from the free-running VCO frequency. Assume symmetrical operation, which implies a capture range of 50 kHz $\times$ 2 = 100 kHz. Once captured, the VCO follows the input to a 200-kHz deviation, implying a lock range of 200 kHz $\times$ 2 = 400 kHz.

560 LIC PLL

The specifications for the 560 PLL IC are provided in Fig. 6-14. A look at the applications listed provides a clue to the versatility of the PLL. In fact, it seems that applications are limited only by user ingenuity. The 560 PLL provides operation for inputs up to at least 15 MHz and typically to 30 MHz at amplitudes from 100 μV to 1 V. It is ideally suited for FM IF amplification and detection at the standard 10.7-MHz IF frequency. A typical FM demodulation circuit is shown in Figure 2 of the specifications. The *tracking filter* shown in Figure 3 has the ability to "follow" an input signal with a fixed bandpass, even as the signal's center frequency is varying. It can, therefore, limit the noise (by limiting bandwidth) while tracking a low-level varying-frequency signal.

6-6 *STEREO DEMODULATION*

FM stereo receivers are identical to standard receivers up to the discriminator output. At this point, however, the discriminator output contains the 30 Hz to 15 kHz (L + R) signal *and* the 19-kHz subcarrier *and* the 23 kHz to 53 kHz (L − R) signal. If a standard receiver is tuned to a stereo station, its discriminator output may contain the additional frequencies, but even the 19 kHz subcarrier is above the normal audible range, and its audio amplifiers and speaker would probably not pass it anyway. Thus, the standard receiver reproduces the 30 Hz to 15 kHz (L + R) signal (a full monophonic broadcast) and is not affected by the other frequencies. This effect is illustrated in Fig. 6-15(a).

The stereo receiver block diagram in Fig. 6-15(b) becomes more complex after the discriminator. At this point, the three components are separated by filtering action. The (L + R) signal is obtained through a low-pass filter and given a delay so that it reaches the matrix network in step with the (L − R) signal. A 23- to 53-kHz bandpass filter "selects" the (L − R) double-side-band signal. A

DESCRIPTION

The NE560B Phase Locked Loop (PLL) is a monolithic signal conditioner, and demodulator system comprising a VCO, Phase Comparator, Amplifier and Low Pass Filter, interconnected as shown in the accompanying block diagram. The center frequency of the PLL is determined by the free running frequency (f_0) of the VCO. This VCO frequency is set by an external capacitor and can be fine tuned by an optional Potentiometer. The low pass filter, which determines the capture characteristics of the loop, is formed by the two capacitors and two resistors at the Phase Comparator output.

The PLL system has a set of self biased inputs which can be utilized in either a differential or single ended mode. The VCO output, in differential form, is available for signal conditioning frequency synchronization, multiplication and division applications. Terminals are provided for optional extended control of the tracking range, VCO frequency, and output DC level.

The monolithic signal conditioner-demodulator system is useful over a wide range of frequencies from less than 1 Hz to more than 15 MHz with an adjustable tracking range of ±1% to ±15%.

FEATURES

- **FM DEMODULATION WITHOUT TUNED CIRCUITS**
- **NARROW BANDPASS - TO ± 1% ADJUSTABLE**
- **TRACKING RANGE**
- **EXACT FREQUENCY DUPLICATION IN HIGH**
- **NOISE ENVIRONMENT**
- **WIDE TRACKING RANGE ±15%**
- **HIGH LINEARITY - 1% DISTORTION MAX**
- **FREQUENCY MULTIPLICATION AND DIVISION**
- **THROUGH HARMONIC LOCKING**

APPLICATIONS

TONE DECODERS

FM IF STRIPS

TELEMETRY DECODERS

DATA SYNCHRONIZERS

SIGNAL RECONSTITUTION

SIGNAL GENERATORS

MODEMS

TRACKING FILTERS

SCA RECEIVERS

FSK RECEIVERS

WIDE BAND HIGH LINEARITY DETECTORS

ABSOLUTE MAXIMUM RATINGS

Maximum Operating Voltage	26V
Input Voltage	1V Rms
Storage Temperature	-65 °C to 150 °C
Operating Temperature	0 °C to 70 °C
Power Dissipation	300 mw

Limiting values above which serviceability may be impaired

PIN CONFIGURATION

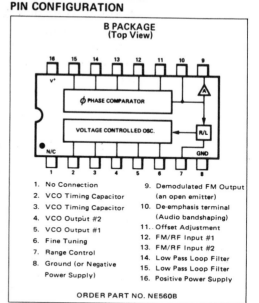

B PACKAGE (Top View)

1. No Connection
2. VCO Timing Capacitor
3. VCO Timing Capacitor
4. VCO Output #2
5. VCO Output #1
6. Fine Tuning
7. Range Control
8. Ground (or Negative Power Supply)
9. Demodulated FM Output (an open emitter)
10. De-emphasis terminal (Audio bandshaping)
11. Offset Adjustment
12. FM/RF Input #1
13. FM/RF Input #2
14. Low Pass Loop Filter
15. Low Pass Loop Filter
16. Positive Power Supply

ORDER PART NO. NE560B

BLOCK DIAGRAM

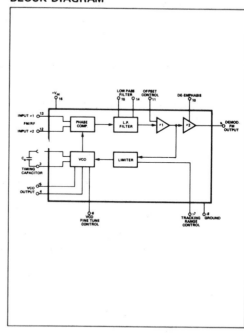

Fig. 6-14. 560 PLL Specifications. (Courtesy of Signetics.)

190

GENERAL ELECTRICAL CHARACTERISTICS

(15KΩ Pin 9 to GND, Input Pin 12 or Pin 13 AC Ground Unused Input, Optional Controls Not Connected, V+ = 18V Unless Otherwise Specified T_A = 25°C)

CHARACTERISTICS	LIMITS			UNITS	TEST CONDITIONS
	MIN	TYP	MAX		
Lowest Practical Operating Frequency		0.1		Hz	
Maximum Operating Frequency	15	30		MHz	
Supply Current	7	9	11	Ma	
Minimum Input Signal for Lock		100		μV	
Dynamic Range		60		dB	
VCO Temp Coefficient*		±0.06	±0.12	%/°C	Measured at 2 MHz, with both inputs AC grounded
VCO Supply Voltage Regulation		±0.3	±2	%/V	Measured at 2 MHz
Input Resistance		2		KΩ	
Input Capacitance		4		Pf	
Input DC Level		+4		V	
Output DC Level	+12	+14	+16	V	
Available Output Swing		4		V_{p-p}	Measured at Pin 9
AM Rejection*	30	40		dB	See Figure 1
De-emphasis Resistance		8		KΩ	

*ACC Test Sub Group C.

ELECTRICAL CHARACTERISTICS (For FM Applications, Figure 2)

(15KΩ Pin 9 to GND, Input Pin 12 or 13, AC Ground Unused Input, Optional Controls Not Connected, V+ = 18V Unless Otherwise Specified T_A = 25°C)

CHARACTERISTICS	LIMITS			UNITS	TEST CONDITIONS
	MIN	TYP	MAX		
10.7 MHz Operation Deviation 75 kHz Source Impedance = 50Ω					
Detection Threshold		120	300	μV	
Demodulated Output Amplitude	30	60		mV	V_{in} = 1 mv Rms Modulation Frequency 1 kHz
Distortion*		.3	1	% T.H.D.	V_{in} = 1 mv Rms Modulation Frequency 1 kHz
Signal to Noise Ratio $\frac{S+N}{N}$		35		dB	V_{in} = 1 mv Rms Modulation Frequency 1 kHz
4.5 MHz Operation Deviation = 25 kHz, Source Impedance = 50Ω					
Detection Threshold		120	300	μV	
Demodulated Output Amplitude	30	60		mV	V_{in} = 1 mv Rms Modulation Frequency 1 kHz
Distortion		0.3	1.0	% T.H.D.	V_{in} = 1 mv Rms Modulation Frequency 1 kHz
Signal to Noise Ratio $\frac{S+N}{N}$		35		dB	V_{in} = 1 mv Rms Modulation Frequency 1 kHz
Wide Deviation $\Delta F/f_o$ = 5% Input = 4.5 MHz Deviation = 225 kHz @ 1 kHz Modulation Rate					
Detection Threshold	0.2	1	5	mV	
Demodulated Output		0.5		Vrms	V_{in} = 5 mv Rms
Distortion		0.8		% T.H.D.	V_{in} = 5 mv Rms
Signal to Noise Ratio $\frac{S+N}{N}$		50		dB	V_{in} = 5 mv Rms

*ACC Test Sub Group C.

ELECTRICAL CHARACTERISTICS (For Tracking Filter, Figure 3)

(15KΩ Pin 9 to GND, Input Pin 12 or Pin 13 AC Ground Unused Input, Optional Controls Not Connected, V+ = 18V Unless Otherwise Specified T_A = 25°C)

CHARACTERISTICS	LIMITS			UNITS	TEST CONDITIONS
	MIN	TYP	MAX		
Tracking Range	±5	±15		% of f_o	V_{in} = 5 mv Rms
Minimum Signal to Sustain Lock 0°C to 70°C		0.8		mv Rms	Input 2 MHz - See Characteristic Curves
VCO Output Impedance		1		kΩ	
VCO Output Swing	0.4	0.6		V_{p-p}	Input 2 MHz Measured with high impedance Probe with less than 10 Pf Capacitance
VCO Output DC Level		+6.5		V	
Side Band Suppression		35		dB	Input 2 MHz with ±100 kHz Side Band Separation and 3 kHz Low Pass Filter Input 1 mv Peak for Carrier Each Side Band C_1 = 0.01 μF R_1 = 0

Fig. 6-14. (Cont.).

191

AM REJECTION

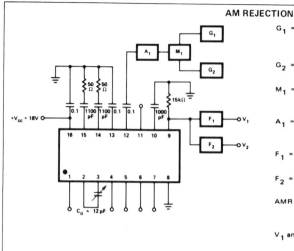

G_1 = FM Generator with f_c = f_0 ≈ 4 MHz
                                         Δf = 40 kHz,
                                         f_{mod} = 1 kHz

G_2 = Audio Generator with f_A = 400 Hz

M_1 = Balanced Modulator Carrier Supplied by G_1, AM
            modulation provided by G_2.

A_1 = 50 Ω attenuator pad with signal level into pin 12 adjusted
            to 1 mV rms.

F_1 = 1 kHz Bandpass filter, Q = 20

F_2 = 400 Hz Bandpass filter with Q = 50, with 1 kHz trap.

$$AMR = \frac{V_1}{V_2} \text{ in dB}$$

V_1 and V_2 are rms voltmeter readings.

Fig. 1

FM DEMODULATION

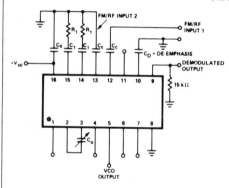

C_B = Bypass Capacitor
C_C = Coupling Capacitors
C_1 = Low Pass Filter Capacitors
C_o = Frequency Determining Capacitor

T_D = De-emphasis time constant
      = (C_D) $(8k\Omega)$

Fig. 2

TRACKING FILTER

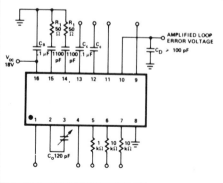

C_C = Coupling Capacitors
C_B = Bypass Capacitor
C_1 = Low Pass Filter Capacitor
C_o = VCO Frequency Set Capacitor

Fig. 3

Fig. 6-14. (Cont.).

MINIMUM INPUT SIGNAL AMPLITUDE NECESSARY TO MAINTAIN LOCK AS A FUNCTION OF TEMPERATURE WITH f_{signal} = fo_{25^oC} = 2.0 MHz

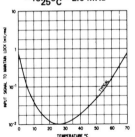

THERMAL DRIFT OF VCO FREE RUNNING FREQUENCY (f_o)

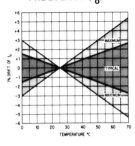

CHANGE OF FREE RUNNING OSCILLATOR FREQUENCY AS A FUNCTION OF FINE TUNING CIRCUIT

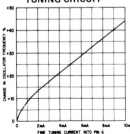

FREE RUNNING OSCILLATOR FREQUENCY AS A FUNCTION OF VCO TIMING CAPACITANCE

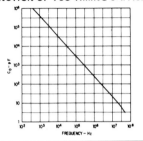

AM REJECTION AS A FUNCTION OF INPUT SIGNAL LEVEL f_o = 10 MHz

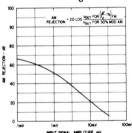

TYPICAL TRACKING RANGE AS A FUNCTION OF INPUT SIGNAL

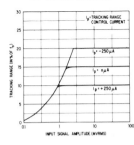

CHANGE OF FREE RUNNING OSCILLATOR FREQUENCY AS A FUNCTION OF RANGE CONTROL CURRENT

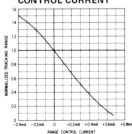

NORMALIZED TRACKING RANGE AS A FUNCTION OF RANGE CONTROL CURRENT

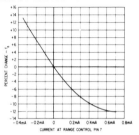

Fig. 6-14. (Cont.).

EXTERNAL CONTROLS

1. Loop Low Pass Filter (Pins 14 and 15)

The equivalent circuit for the loop low-pass filter can be represented as:

where RA (6K Ω) is the effective resistance seen looking into Pin #14 or Pin #15.

The corresponding filter transfer characteristics are:

$$\frac{V_2}{V_1}(S) = (S) = \frac{1 + S R_1 C_1}{1 + S (R_1 + R_A) C_1}$$

where S is the complex frequency variable.

2. Loop Gain (Threshold) Control

The overall Phase Locked Loop gain can be reduced by connecting a feedback resistor, R_F, across the low-pass filter terminals, Pins #14 and #15. This causes the loop gain and the detection sensitivity to decrease by a factor α ($\alpha < 1$) where:

$$\alpha = \frac{R_F}{2 R_A + R_F}$$

Reduction of loop gain may be desirable at high input signal levels ($V_{in} > 30$ mV) and at high frequencies ($f_0 > 5$ MHz) where excessively high loop gain may cause instability.

3. Tracking Range Control (Pin 7)

Any bias current, I_p, injected into the tracking range control, reduces the tracking range of the PLL by decreasing the output of the limiter. The variation of the tracking range and the center frequency, as a function of I_p, are shown in the characteristic curves with I_p defined positive going into the tracking range control terminal. This terminal is normally at a DC level of +0.6 Volts and presents an impedance of 600 Ω.

4. External Fine Tuning (Pin 6)

Any bias current injected into the fine tuning terminal increases the frequency of oscillation, f_0, as shown in the characteristic curves. This current is defined Positive into the fine tuning terminal. This terminal is at a typical DC level of +1.3 Volts and has a dynamic impedance of 100Ω to ground.

5. Offset Adjustment (Pin 11)

Application of a bias voltage to the offset adjustment terminal modifies the current in the output amplifier setting the DC level at the output. The effect on the loop is to modify the relationship between the VCO free running frequency and the lock range, allowing the VCO free running frequency to be positioned at different points throughout the lock range.

Nominally this terminal is at +4V DC and has an input impedance of 3K Ω. The offset adjustment is optional. The characteristics specified correspond to operation of the circuit with this terminal open circuited.

6. De-emphasis Filter (Pin 10)

The de-emphasis terminal is normally used when the PLL is used to demodulate Frequency Modulated Audio signals. In this application, a capacitor from this terminal to ground provides the required de-emphasis. For other applications, this terminal may be used for band shaping the output signal. The 3 dB bandwidth of the output amplifier in the system block diagram (see Figure 2 .) is related to the de-emphasis capacitor, C_D, as:

$$f_{3dB} = \frac{1}{2 \, R_a C_D}$$

where R_D is the 8000 ohm resistance seen looking into the de-emphasis terminal.

When the PLL system is utilized for signal conditioning, and the loop error voltage is not utilized, de-emphasis terminal should be AC grounded.

Fig. 6-14. (Cont.).

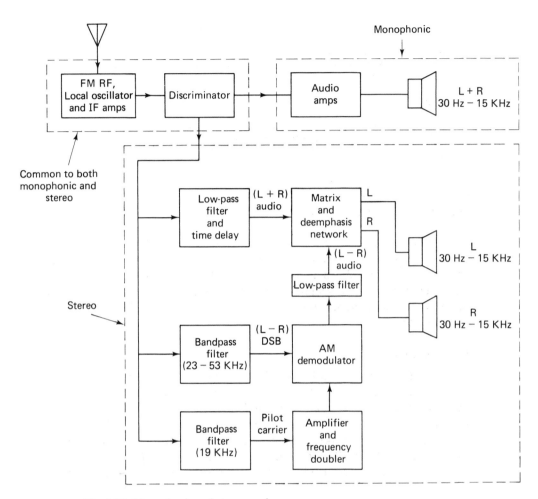

Fig. 6-15. Monophonic and stereo receivers.

19-kHz bandpass filter takes the pilot carrier that is then multiplied by 2 to 38 kHz, which is the precise carrier frequency of the DSB suppressed carrier 23 kHz to 53 kHz (L − R) signal. Combining the 38 kHz and (L − R) signals through the nonlinear device of an AM detector generates sum and difference outputs of which the 30 Hz to 15 kHz (L − R) components are selected by a low-pass filter. The (L − R) signal is thereby retranslated back down to the audio range and it and the (L + R) signal are applied to the matrix network for further processing. Figure 6-16 illustrates the matrix function and completes the stereo receiver block diagram of Fig. 6-15(b). The (L + R) and (L − R) signals are combined in an adder that cancels R since (L + R) + (L − R) = 2L. The L − R signal is also applied to an inverter, providing −(L − R) = (−L + R), which is subsequently applied to

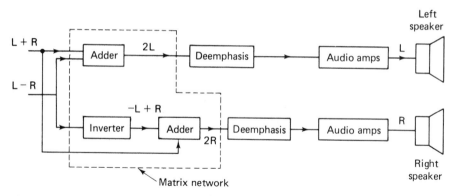

Fig. 6-16. Stereo signal processing.

another adder along with (L + R), which produces (−L + R) + (L + R) = 2R. The two individual signals for the right and left channels are then deemphasized and individually amplified to their own speaker. The process of FM stereo is ingenious in its relative simplicity and effectiveness in providing complete compatibility and doubling the amount of transmitted information through the use of multiplexing.

SCA Decoder

The FCC has also authorized FM stations to broadcast background music programming which is usually commercial-free but paid for by subscription of department stores, groceries, and the like. It is termed the *subsidiary communication authorization* (SCA). It is also frequency-multiplexed on the FM modulating signal, usually with a 67-kHz carrier and ±7.5 kHz (narrow-band) deviation, as shown in Fig. 6-17. An SCA decoder circuit using the 565 PLL is provided in Fig. 6-18. A resistive voltage divider is used to establish a bias voltage for the input

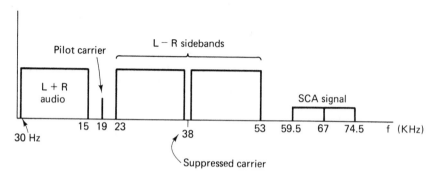

Fig. 6-17. Composite stereo and SCA modulating signal.

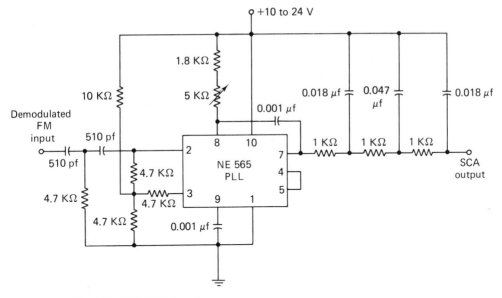

Fig. 6-18. SCA PLL decoder.

(pins 2 and 3). The demodulated FM signal is fed to the input through a two-stage high-pass filter (510 pF, 4.7 kΩ, 510 pF, 4.7 kΩ), both to effect capacitive coupling and to attenuate the stronger level of the stereo signals. The PLL is tuned to about 67 kHz, with the 0.001-μF capacitor from pin 9 to ground with the 5-kΩ potentiometer providing fine adjustment. The demodulated output at pin 7 is fed through a three-stage low-pass filter to provide deemphasis and attenuate the high-frequency noise which often accompanies SCA transmission.

LIC Stereo Decoder

The decoding of the stereo signals is normally accomplished via special function ICs. The RCA CA3090 is such a device with a functional block diagram provided in Fig. 6-19. It is a very complex IC containing hundreds of transistors but fortunately is easy to use. The input signal from the detector is amplified by a low-distortion preamplifier and simultaneously applied to both the 19 kHz and 38 kHz synchronous detectors. A 76-kHz signal, generated by a local voltage-controlled oscillator (VCO), is counted down by two frequency dividers to a 38-kHz signal and to two 19-kHz signals in phase quadrature. The 19-kHz pilot tone supplied by the FM detector is compared to the locally generated 19-kHz signal in the synchronous detector. The resultant signal controls the voltage-controlled oscillator so that it produces an output signal to phase-lock the stereo decoder with the pilot tone. A second synchronous detector compares the locally

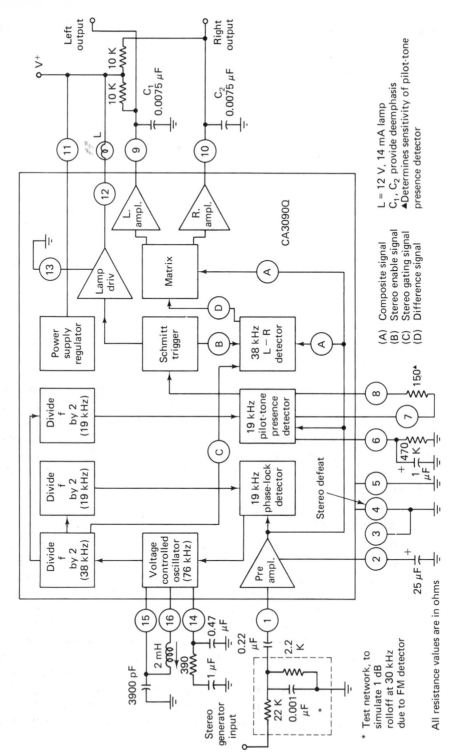

Fig. 6-19. CA3090 stereo decoder. (Courtesy of RCA.)

L = 12 V, 14 mA lamp
C_1, C_2 provide deemphasis
▲Determines sensitivity of pilot-tone presence detector

(A) Composite signal
(B) Stereo enable signal
(C) Stereo gating signal
(D) Difference signal

All resistance values are in ohms

* Test network, to simulate 1 dB rolloff at 30 kHz due to FM detector

Stereo generator input

Stereo defeat

CA3090Q

198

generated 19-kHz signal with the 19-kHz pilot tone. If the pilot tone exceeds an externally adjustable threshold voltage, a Schmitt trigger circuit is energized. The signal from the Schmitt trigger lights the stereo indicator, enables the 38-kHz synchronous detector, and automatically switches the CA3090 from monaural to stereo operation. The output signal from the 38-kHz detector and the composite signal from the preamplifier are applied to a matrixing circuit, from which emerge the resultant left and right channel audio signals. These signals are applied to their respective left and right channel amplifiers for amplification to a level sufficient to drive most audio power amplifiers.

6-7 FM RECEIVERS

The current state-of-the-art for FM receivers involves use of discrete MOSFET RF and mixer stages with a separately excited bipolar transistor local oscillator as shown in Fig. 6-20. The antenna input signal is applied through the tuning circuit L_1, C_{1A} to the gate of the 40822 MOSFET RF amplifier. Its output at the drain is coupled to the lower gate of the 40823 mixer MOSFET through the C_{1B}, L_2 tuned circuit. The 40244 BJT oscillator signal is applied to the upper gate of the mixer stage. The local oscillator tuned circuit, L_3 and C_{1C}, uses a tapped inductor indicating a Hartley oscillator configuration. The tuning condensor, C_1, has three separate ganged capacitors which vary the tuning range of the RF amp and mixer tuned circuits from 88 to 108 MHz while varying the local oscillator frequency from 98.7 to 118.7 MHz to generate a 10.7-MHz IF signal at the output of the mixer. The mixer output is applied to the commercially available 10.7-MHz double-tuned circuit T_1.

MOSFET receiver front ends offer superior cross-modulation and inter-modulation performance as compared to BJT or vacuum-tube circuits, as previously explained in Sec. (6-2). The *Institute of High Fidelity Manufacturers* (IHFM) sensitivity for this front end is about 1.75 μV. It is defined as the minimum 100% modulated input signal that reduces the total receiver noise and distortion to 30 dB below the output signal. In other words a 1.75-μV input signal produces 30-dB *quieting*.

The front-end output through T_1 in Fig. 6-20 is applied to a CA3089 IC. The CA3089 provides three stages of IF amplification, limiting, demodulation, and audio preamplification. It also provides a signal to drive a tuning meter and an AFC output for direct control of a varactor tuner. It is a most versatile, low-cost, circuit element. Its audio output includes the 30 Hz to 15 kHz (L + R) signal, 19-kHz pilot carrier, and 23 kHz to 53 kHz (L − R) signal, which are then applied to the FM stereo decoder IC, the CA3090. The CA3090 was explained in the previous section and provides the separated left and right channel outputs as well as a signal to light a stereo indicator light.

The CA3090 audio outputs are then applied to a dual ganged volume control

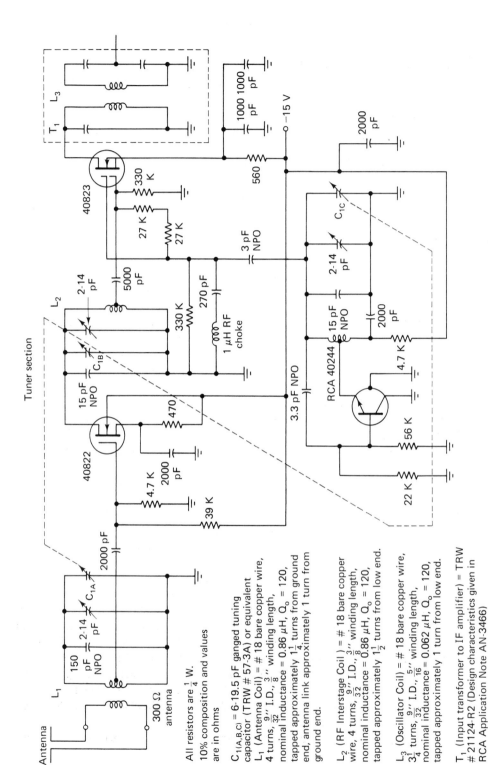

Fig. 6-20. Complete 88 MHz to 108 MHz stereo receiver.

Tuner section

Antenna

300 Ω antenna

L_1

All resistors are $\frac{1}{4}$ W. 10% composition and values are in ohms

$C_{1(A,B,C)}$ = 6-19.5 pF ganged tuning capacitor (TRW # 57-3A) or equivalent
L_1 (Antenna Coil) = # 18 bare copper wire, 4 turns, $\frac{9}{32}$'' I.D., $\frac{3}{8}$'' winding length, nominal inductance = 0.86 μH, Q_o = 120, tapped approximately $1\frac{1}{4}$ turns from ground end, antenna link approximately 1 turn from ground end.

L_2 (RF Interstage Coil) = # 18 bare copper wire, 4 turns, $\frac{9}{32}$'' I.D., $\frac{3}{8}$'' winding length, nominal inductance = 0.86 μH, Q_o = 120, tapped approximately $1\frac{1}{2}$ turns from low end.

L_3 (Oscillator Coil) = # 18 bare copper wire, $3\frac{1}{4}$ turns, $\frac{9}{32}$'' I.D., $\frac{5}{16}$'' winding length, nominal inductance = 0.062 μH, Q_o = 120, tapped approximately 1 turn from low end.

T_1 (Input transformer to IF amplifier) = TRW # 21124-R2 (Design characteristics given in RCA Application Note AN-3466)

200

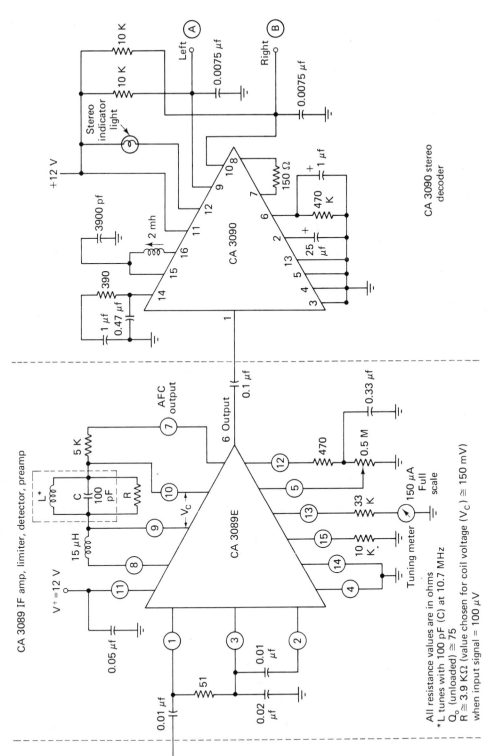

Fig. 6-20. (Cont.).

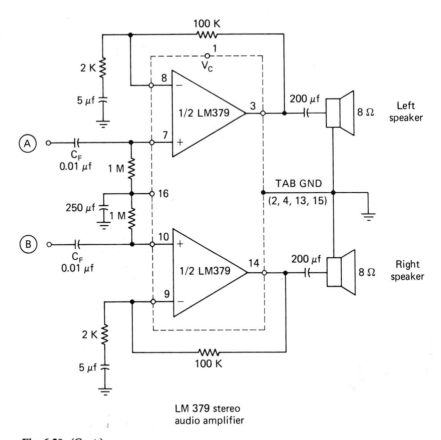

LM 379 stereo
audio amplifier

Fig. 6-20. (Cont.).

potentiometer and then to a LM379 dual 6-W audio amplifier. It has two separate audio amplifiers in one 16 lead IC and has a minimum input impedance of 2 MΩ per channel. It typically provides a voltage gain of 34 dB, *total harmonic distortion* (THD) of 0.07% at 1 W output, and 70 dB of channel separation.

7
COMMUNICATION
TECHNIQUES

Communications equipment may be defined loosely as that which is *not* used for entertainment. Since much of this equipment is of a vital nature, it is not surprising that communications equipment is more sophisticated than a standard broadcast receiver. In addition, since communications tends to require two-way capabilities, we find the use of transceivers prevalent. A *transceiver* is simply a transmitter *and* receiver in a single package.

7-1 FREQUENCY CONVERSION

Double Conversion

One of the most likely areas of change from broadcast receivers to communications receivers is in the mixing process. The two major differences are the widespread use of double conversion and the increasing popularity of up-conversion in communication equipment. Both of these refinements have as a major goal the minimization of image frequency problems. (Refer to Chapter 3 for a review of these phenomena).

Double conversion is the process of first stepping down the signal to a first, relatively high, IF frequency and then mixing again down to a second, lower, final IF frequency. Figure 7-1 provides a block diagram for a typical double-conversion system. Note that the first local oscillator is variable so as to allow a constant 10-MHz frequency for the first IF amplifier. Now the input into the second mixer is a constant 10 MHz, which allows the second local oscillator to be a fixed 11-MHz crystal oscillator. The difference component (11 MHz − 10 MHz = 1 MHz) out of the second mixer is acceptable for the second IF amplifier, which is operating at the comfortable low frequency of 1 MHz. The following example will serve to illustrate the ability of double conversion to eliminate image frequency problems.

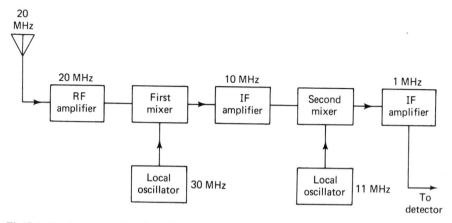

Fig. 7-1. Double conversion block diagram.

EXAMPLE 7-1

Determine the image frequency for the receiver illustrated in Fig. 7-1.

Solution:

The image frequency is the one that when mixed with the 30-MHz first local oscillator signal will produce a first mixer output frequency of 10 MHz. The desired frequency of 20 MHz mixed with 30 MHz yields a 10-MHz component, of course, but what *other* frequency provides a 10-MHz output? A little thought shows that if a 40-MHz input signal mixes with a 30 MHz local oscillator signal, an output of 40 MHz − 30 MHz = 10 MHz is also produced. Thus, the image frequency is 40 MHz.

Example 7-1 shows that in this case the image frequency is double the desired signal (40 MHz vs. 20 MHz), and even the relatively broad-band tuned circuits of the RF and mixer stages will almost totally suppress the image frequency. On the other hand, if this receiver used a single conversion directly to the final

1 MHz IF frequency, it is found that the image frequency will not be fully suppressed.

EXAMPLE 7-2

Determine the image frequency for the receiver illustrated in Fig. 7-2.

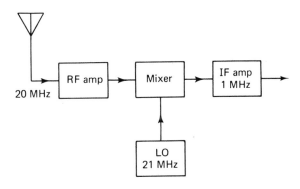

Fig. 7-2. System for Example 7-2.

Solution:

If a 22-MHz signal mixes with the 21-MHz local oscillator, a difference component of 1 MHz is produced just as when the desired 20-MHz signal mixes with 21 MHz. Thus, the image frequency is 22 MHz.

The 22-MHz image frequency of Ex. 7-2 is very close to the desired 20-MHz signal. While the RF and mixer tuned circuits will certainly provide attenuation to the 22-MHz image, if it is a strong signal, it will certainly get into the IF stages and will not be removed from that point on. The graph of RF and mixer tuned circuit response in Fig. 7-3 serves to illustrate the tremendous image frequency response rejection provided by the double-conversion scheme.

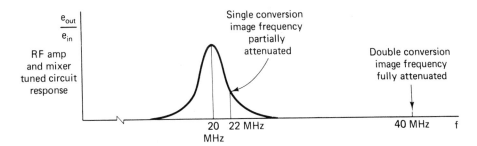

Fig. 7-3. Image frequency rejection.

Image frequencies are not a major problem for low-frequency carriers—say, below 4 MHz. For example, a single-conversion setup for a 4-MHz carrier and a 1-MHz IF means that a 5-MHz local oscillator will be used. The image frequency is 6 MHz, which is far enough away from the 4-MHz carrier so as to not present a problem. At higher frequencies, where images are a problem, as shown in Ex. 7-2, the situation is aggravated by the enormous number of transmissions taking place in our crowded communications bands.

EXAMPLE 7-3

Why do you suppose that images tend to be somewhat less of a problem in FM versus AM or SSB communications?

Solution:

Recall the concept of the *capture* effect in FM systems (Chapter 5). It was shown that if a desired and an undesired station are picked up simultaneously, the stronger one tends to be "captured" by inherent suppression of the weaker signal. Thus, a 2:1 signal to undesired signal ratio may result in a 10:1 ratio at the output. This contrasts with AM systems (SSB included) where the 2:1 ratio is carried through right to the output.

Up-Conversion

Until recently, the double-conversion scheme, with the lowest IF frequency (often the familiar 455 kHz) doing most of the receiver's selectivity, has been standard practice. This was because the components available made it easiest to achieve the necessary selectivity at low IF frequencies. However, now that VHF crystal filters (30 to 120 MHz) are available for IF circuitry, conversion to a higher IF than RF frequency is becoming popular in sophisticated communications receivers. As an example, consider a receiver tuned to a 30-MHz station and using a 40-MHz IF frequency as illustrated in Fig. 7-4. This represents an *up-conversion*

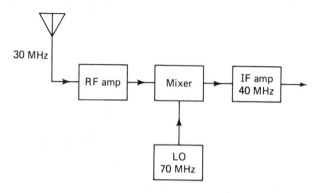

Fig. 7-4. An up-conversion system.

system in that the IF is a higher frequency than the received signal. The 70-MHz local oscillator mixes with the 30-MHz signal to produce the desired 40-MHz IF. Sufficient IF selectivity at 40 MHz is possible with a crystal filter.

EXAMPLE 7-4

Determine the image frequency for the system of Fig. 7-4.

Solution:

If a 110-MHz signal mixes with a 70-MHz local oscillator, a 40-MHz output component results. The image frequency is, therefore, 110 MHz.

The preceding example shows the superiority of up-conversion. It is highly unlikely that the 110-MHz image could get through the RF amplifier tuned to 30 MHz. There is no need for double conversion and all its necessary extra circuitry. The only disadvantage to up-conversion is the need for a high-Q IF filter and better high-frequency-response IF transistors. The current state-of-the-art in these areas now makes up-conversion economically attractive. Additional advantages over double conversion include better image suppression and fewer tuning range requirements for the oscillator. The smaller tuning range for up-conversion is illustrated in the following example, and it minimizes the tracking problems of the receiver.

EXAMPLE 7-5

Determine the local oscillator tuning range for the systems illustrated in Figs. 7-1 and 7-4 if the receivers must tune from 20 to 30 MHz.

Solution:

The double-conversion local oscillator in Fig. 7-1 is at 30 MHz for a received 20-MHz signal. To provide the same 10-MHz IF frequency for a 30-MHz signal means that the local oscillator must be at 40 MHz. Its tuning range is from 30 to 40 MHz or 40 MHz/30 MHz = 1.33. The up-conversion scheme of Fig. 7-4 has a 70-MHz local oscillator for a 30-MHz input and requires an 80-MHz oscillator for a 20-MHz input. Its tuning ratio is then 80 MHz/70 MHz, or a very desirably low value of 1.14.

7-2 SPECIAL FEATURES

Delayed AGC

The simple automatic gain control (AGC) discussed in Chapter 3 has a minor disadvantage. It provides some gain reduction even to very weak signals. This is illustrated in Fig. 7-5. As soon as even a weak received signal is tuned, simple AGC provides some gain reduction. Since communication equipment is

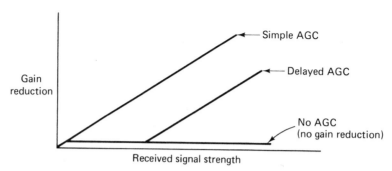

Fig. 7-5. AGC characteristics.

often dealing with marginal (weak) signals, it is usually advantageous to provide some additional circuitry to provide a *delayed* AGC: that is, an AGC that does not provide any gain reduction until some arbitrary signal level is attained and, therefore, has no gain reduction for weak signals. This characteristic is also shown in Fig. 7-5.

A simple means of providing delayed AGC is shown in Fig. 7-6. A reverse bias is applied to the cathode of CR_1. Thus, the diode looks like an open circuit to the ac signal from the last IF amplifier unless that signal reaches some predetermined instantaneous level. For small IF outputs when CR_1 is an open, the capacitor C_1 sees a pure ac signal and thus no dc AGC level is sent back to previous stages to reduce their gain. If the IF output increases, eventually a point is reached where CR_1 will conduct on its peak positive levels. This will effectively "short" out the positive peaks of IF output and C_1 will, therefore, see a more negative than positive signal and filter it into a relatively constant negative level used to reduce the gain of previous stages. The amplitude of IF output required to start feedback of the "delayed" AGC signal is adjustable by the delayed AGC control potentiometer of Fig. 7-6. This may be an external control so that the user can adjust the amount of delay to suit conditions. For instance, if mostly weak signals are being received, the control might be set so that no AGC signal is developed except for very strong stations. This means that the delay interval shown in Fig. 7-5 is increased.

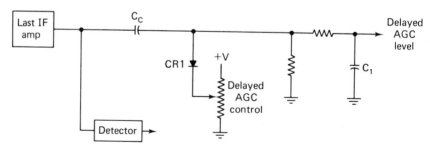

Fig. 7-6. Delayed AGC configuration.

Auxiliary AGC

Auxiliary AGC is used (even on some broadcast receivers) to cause a step reduction in receiver gain at some arbitrarily high value of received signal. It then has the effect of preventing very strong signals from overloading a receiver. A simple means of accomplishing the auxiliary AGC function is illustrated in Fig. 7-7. Notice the auxiliary AGC diode connected between the collectors of the mixer and first IF transistors. Under normal signal conditions, the dc level at each collector is such that the diode is reverse-biased. In this condition, the diode has a very high resistance and has no effect on circuit action. The potential at the mixer's collector is constant since it is not controlled by the normal AGC. However, the AGC control on the first IF transistor, for very strong signals, causes its dc base current to increase, and hence the collector current also increases. Thus, its collector voltage becomes less negative and the diode starts to conduct. The diode resistance goes low and it loads down the mixer tank (L_1C_1) and, thereby, produces a step reduction of the signal coupled into the first IF stage. The dynamic AGC range has thereby been substantially increased.

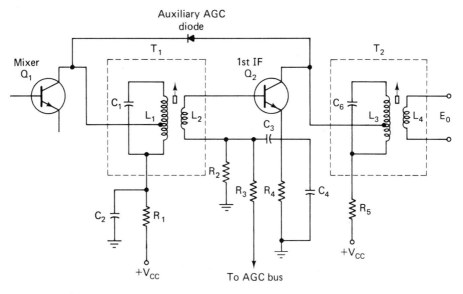

Fig. 7-7. Auxiliary AGC.

Band Spread

Communications receivers are often called upon to pick out a station transmitting on a narrow frequency band in close proximity to a number of other transmissions. In those cases it is difficult to tune the station unless some form of band spread is incorporated. *Band spreading* is the spreading of tuning indications over a wide scale range to facilitate tuning in a crowded band of frequencies. Either electrical or mechanical means are used to provide band spread.

Mechanical means are used to provide a second calibrated tuning dial that is "geared" down from the main dial such that a full revolution is only a small movement of the main dial. This is similar to a vernier control. Electrical band spread can be accomplished through use of a second, but smaller, tuning capacitor in parallel with the main tuning capacitor. The need for either type of band spread is diminishing as the use of frequency-synthesized communications equipment becomes more prevalent. In synthesized receivers, the desired station is accurately selected by "dialing in" or pushbutton selection of the necessary local oscillator frequency.

Variable Sensitivity

In spite of the increased dynamic range provided by delayed AGC and auxiliary AGC, it is often advantageous for a receiver to also include a variable sensitivity control. This is a manual AGC control in that the user controls the receiver gain (and thus sensitivity) to suit the requirement. A communications receiver may be called upon to deal with signals over a 100,000:1 ratio, and even the most advanced AGC system does not afford that amount of range. Receivers that are designed to provide high sensitivity and also are able to handle very large input signals incorporate a manual sensitivity control.

Variable Selectivity

Many communications receivers provide detection to more than one kind of transmission. They may detect code transmissions, SSB, AM, and FM all in one receiver. The required bandwidth to avoid picking up adjacent channels may well vary from 1 kHz for code up to 20 kHz for FM. Variable selectivity may be accomplished by having a potentiometer across the primary and/or secondary of the last IF stage. The variable resistance controls the Q and hence the BW of the tuned circuit. This control can also be advantageous in situations where a received signal with a 10-kHz bandwidth is being adversely affected by noise. By reducing the receiver's selectivity to only 5 kHz, the external noise should be roughly cut in half, making a marginal reception acceptable. This action will, however, cut off some of the received side bands and, therefore, decrease some of the intelligibility. The reduced selectivity therefore is a compromise adjustment in this situation.

Noise Limiter

Man-made sources of external noise are extremely troublesome to highly sensitive communications receivers, as well as to any other electronics equipment that is dealing with signals in the microvolt region or less. The interference created by these man-made sources, such as ignition systems, motor commutation systems, and switching of high-current loads, is a form of *electromagnetic interference*

(EMI). Reception of EMI by a receiver creates undesired amplitude modulation, sometimes of such large magnitude as to adversely affect FM reception, to say nothing of the complete havoc created in AM systems. While these noise impulses are usually of short duration, it is not uncommon for them to have amplitudes up to 1000 times that of the desired signal! A noise limiter circuit is employed to silence the receiver for the duration of a noise pulse, which is preferable to a very loud crash from the speaker. These circuits are sometimes referred to as *automatic noise limiter* (ANL) *circuits.*

A common type of circuit for providing noise limiting is shown in Fig. 7-8.

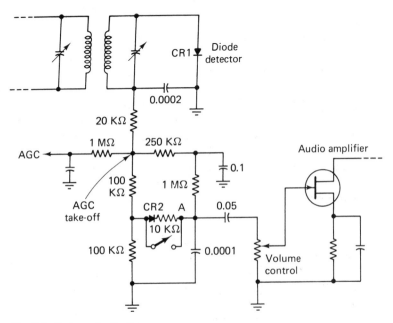

Fig. 7-8. Series noise limiter.

It uses a diode, CR_2, that conducts the detected signal to the audio amplifier as long as the signal is not greater than some prescribed limit. Greater amplitudes cause CR_2 to stop conducting until the noise impulse has decreased or ended. The varying audio signal from the diode detector, CR_1, is developed across the two 100-kΩ resistors. If the received carrier is producing a -10-V level at the AGC takeoff, the anode of CR_2 is at -5 V and the cathode is at -10 V. The diode is "on" and conducts the audio into the audio amplifier. Impulse noise will cause the AGC takeoff voltage to instantaneously increase, which means that the anode of CR_2 does also. However, its cathode potential does not change instantaneously, since the voltage from cathode to ground is across a 0.001-μF capacitor. Remember that the voltage across a capacitance cannot instantaneously change. Therefore, the cathode stays at -10 V and as the anode approaches -10 V, CR_2 turns "off" and the detected audio is blocked from entering the audio amplifier. The receiver is silenced for the duration of the noise pulse.

The switch across CR_2 allows the noise limiter action to be disabled by the user. This may be necessary when a marginal (noisy) signal is being received and the set output is being turned off excessively by the ANL.

Metering

Most communications receivers are equipped with a meter that provides a visual indication of received signal strength. It is known as the *S meter* and often is found in the emitter leg of an AGC controlled amplifier stage (RF or IF). It reads dc current, which is usually inversely proportional to received signal strength. With no received signal, there is no AGC bias level which causes maximum dc emitter current flow and therefore maximum stage voltage gain. As the AGC level increases, indicating an increasing received signal, the dc emitter current goes down, which thereby reduces gain. The *S* meter can thus be used as an aid to accurate tuning as well as providing a relative guide to signal strength.

In some receivers, the *S* meter can be electrically switched into different areas so as to allow its use in troubleshooting a malfunction. In those cases, the operator's manual provides a troubleshooting guide on the basis of current readings obtained in different areas of the receiver.

Squelch

When a sensitive receiver receives no carrier, the AGC action causes maximum system gain, which results in a high degree of noise output. This occurs in many communication applications where the user is constantly monitoring a transmission, such as in police service, but there is no transmission most of the time. Without a squelch system to cut off the receiver's output during transmission lulls, the noise output would cause the user severe aggravation. Squelch circuitry is also useful in minimizing noise output that occurs when tuning between stations. In fact, even better quality broadcast FM receivers provide a squelch capability, but in these applications, it is usually termed *muting*. Squelch circuitry is also referred to as the *quieting* or *Q circuit*.

Figure 7-9 shows a squelch configuration that causes the audio amplifier stage to be cut off whenever no carrier (or an extremely weak station) is being received. In that case, the AGC level is zero, which causes the dc amplifier stage, Q_2, to be on. This draws the current being supplied by R_1 into Q_2's collector away from Q_1's base, which means that the audio preamp transistor, Q_1, is cut off. Thus, no ac signal is available from Q_1's collector for subsequent power amplification to the speaker. When a signal is picked up by the receiver, the AGC level goes to some negative dc value, which causes Q_2 to turn off. This allows the dc current being supplied by R_1 to enter Q_1's base, which biases it on so that audio preamplification can take place.

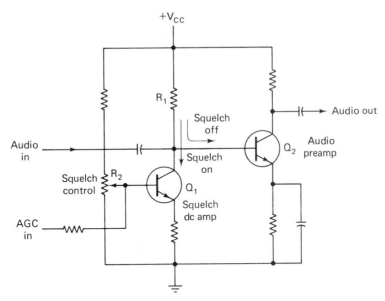

Fig. 7-9. Squelch circuit.

User adjustment of the squelch control (R_2) allows the cut-in point of quieting to be varied. This is necessary so that a very weak station, one that generates a small AGC level, will not cause the receiver's output to be squelched.

7-3 CB TRANSCEIVERS

In recent years the first widespread use, by the average man on the street, of communications equipment has occurred. The mushrooming growth of *citizens' band* (CB) has been most startling. CB was established in 1958 by the FCC for personal or business communications. They set up 23 channels in the 26.965- to 27.255-MHz region. Transmitted powers of 4 W for AM and 12 W of peak envelope power (pep) for SSB operation are the maximum allowed limits. They result in a transmission range of about 25 miles in rural areas down to perhaps 3 miles in noisy urban areas. SSB operation provides increased range over AM, as explained in Sec. (4-1). Operators are required to obtain a license, but there are no test requirements.

The CB boom started during the gasoline shortage of 1973–1974. Truck drivers bought sets to help them find the location of their next source of gasoline. Then the government lowered speed limits to conserve gas and the radios were used to provide warnings of the locations of "Smokeys" or "bears," which is CB slang for state troopers. At this point, the general public joined the bandwagon such that the 1 million people licensed from 1958 to 1974 were joined by another

5 million by 1976! This popularity strained the 23 allocated channels to the point that the FCC expanded to 17 additional channels (40 total) in 1977.

To gain this mass appeal, CB transceivers have had to offer a relatively low purchase price of $50 to $200 and perhaps an additional $100 for the SSB units, with their resultant complexities.

Typical CB Transceiver

A typical mobile CB unit is pictured in Fig. 7-10. It is the Dynascan Cobra Model 29 unit. It is powered directly from the 12 V dc vehicle electrical system which is provided regulation by the transceiver. The receiver RF section has an external gain control which further augments the AGC range of 1 μV (its sensitivity) up to 10,000 μV. It uses double conversion to minimize images. The incoming signal is converted down to 11.275 MHz by a 38.240- to 38.530-MHz local oscillator signal produced by a crystal quasi-synthesizer. It is termed "quasi" in that a bank of 10 crystals is used to provide the 23 different receive local oscillator frequencies and the 23 required transmit frequencies. This is better than the 46 crystals required otherwise (for a 23-channel transceiver) but not as good as the one or two crystals required for a fully synthesized transceiver. The Cobra synthesizer produces the sum frequency of one of six crystals in the 23.290- to 23.540-MHz range and one of four crystals in the 14.950- to 14.990-MHz range by heterodyning the two. Mixing the first IF of 11.275 MHz with a crystal-controlled 11.730-MHz signal creates the 455-kHz second IF. Sharp selectivity is obtained at

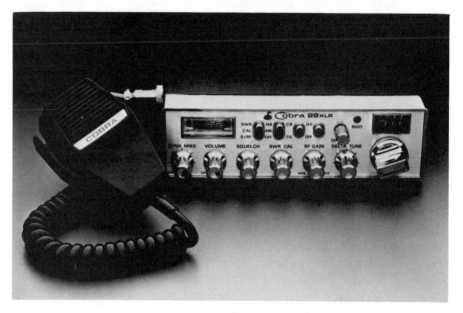

Fig. 7-10. CB transceiver. (Courtesy of Dynascan Corp.)

this point through use of a 455-kHz ceramic filter. The squelch control provides a threshold range of from 0.35 μV up to 250 μV. This means that receiver output can be cut off for received signals over that range at the user's discretion. An *automatic noise limiter* (ANL) can be switched in or out. The ANL circuit is in the RF amp, cutting out any very short duration noise spikes. It works so fast that you do not "miss" any of the desired signal. This augments the unit's standard noise limiter [see Sec. (7-2)], which shuts off the audio amp for longer-duration noise signals.

The transmit section of the CB rig shown in Fig. 7-10 is of standard design. The output of the synthesizer is mixed with an 11.275-MHz crystal signal to produce the carrier. The driver and output stages are collector-modulated and a sample of the modulating signal is fed back to a separate microphone amplifier stage, where it is used to provide an automatic modulation control. This prevents overmodulation ($>100\%$), which is prohibited by the FCC because of the spurious interfering frequencies thereby produced. A multisection network in the power amplifier stage provides impedance matching into the antenna. A television interference (TVI) trap is also provided to ensure that spurious outputs do not interfere with TV receivers. The second and third harmonics of the carrier fall in the VHF TV band. An electronic transmit/receive (TR) switch converts the transceiver from receive to transmit functions when the "push-to-talk" microphone button is activated. If you have a need for public address capability, switching the CB/PA switch to PA allows you to talk into the microphone with power amplification to the set's speaker. The meter on the transceiver's face provides an indication of transmitted power during transmit and of received signal strength during reception.

The Cobra transceiver also includes the *delta tune* feature. This allows the user to tune the receiving frequency slightly off center-channel to compensate for variations in the transmitting frequency. The FCC requires carrier accuracies of 0.005% (50 ppm), and this amount of variation between transmitter to receiver can be annoying without the delta tune adjustment.

CB Frequency Synthesizers

There is a strong trend toward total usage of special-purpose ICs for CB transceivers. This applies to the usage of special LSI chips to accomplish true frequency synthesis. The Nitron NC6400 chip provides the entire decoding and controlling circuitry for a PLL frequency synthesizer. The block diagram in Fig. 7-11 incorporates the NC6400 to accomplish an 80-channel AM CB transceiver with a single crystal. The 80 distinct channels can be called up by keying in two digits and an enter command on a standard three-by-four matrix keyboard such as used in touch-tone telephones, or by serially entering two binary coded (BCD) digits on a four-line data bus. A transmit/receive input allows the user to choose either of two frequencies for each channel. The particular data input coding and corresponding synthesized frequencies can be programmed into the read-only memory (ROM) during the NC600s manufacturing process.

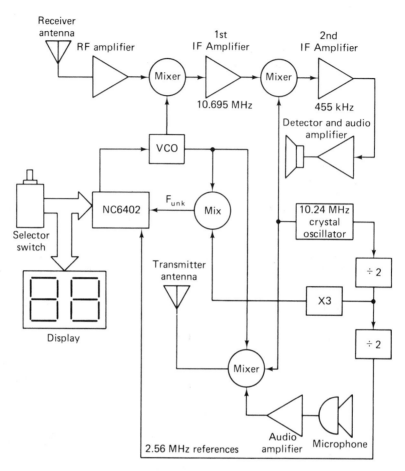

Fig. 7-11. PLL CB frequency synthesizer.

7-4 FACSIMILE

Facsimile is the process whereby fixed graphic material, such as pictures, drawings, or written material, is scanned and converted to an electrical signal, transmitted, and after reception used to produce a likeness (facsimile) of the subject copy. It is roughly comparable to the transmission of a single TV frame, except that output is reproduced on paper rather than the face of a CRT. Facsimile has been used for years for the rapid transmission of photos to local newspapers by news services (e.g., AP wirephoto) and aboard ships for reception of up-to-date weather charts or maps.

In recent years, facsimile equipment has been designed to appeal to the industrial world. This equipment generally uses standard telephone lines as the communication link and is acoustically coupled by simply placing a telephone handset into a special coupler. This enables industry to rapidly send important business papers across town or around the world using standard telephone lines. As the cost and problems concerned with standard mail delivery escalate, it is not far-fetched to envision facsimile postal service.

A representative example of a current facsimile transceiver is the Xerox telecopier 200 shown in Fig. 7-12. This unit consists of a scan/print processor mounted on top of a pedestal containing the system's electronics. The copy to be transmitted is scanned by an 8-mW helium–neon laser light source. Light reflected from the light and dark elements on each page is gathered by a solid-state photoreceptor. This signal is then used to frequency-modulate a low-frequency carrier before it is applied to the phone line.

Fig. 7-12. Xerox telecopier 200 facsimile transceiver (Courtesy of Xerox Corp.)

Standard phone lines have a very limited bandwidth of less than 3 kHz. Recall Hartley's law (Chapter 1), which states that the information transmitted is proportional to bandwidth times the time of transmission. Because of the limited bandwidth, it takes 2 minutes for the Telecopier 200 to transmit the information on an 8-1/2- by 11-inch sheet of paper. This is in sharp contrast to our television system, where one picture is transmitted in $\frac{1}{30}$ second *but* using a bandwidth of 6 MHz.

In the receive mode, the Telecopier 200 demodulates the information and applies it to the same laser beam used for scanning, whose intensity now varies with the information. The modulated beam is then used to reproduce the original copy using the same xerographic technology used in standard Xerox copiers.

7-5 MOBILE TELEPHONE

Mobile telephone service is designed to connect mobile units with dial telephone exchanges. The more advanced systems provide fully automatic operation, giving the mobile user identical telephone capabilities as a regular fixed subscriber. Earlier systems require the mobile user to manually place a call through an operator. Each mobile unit is assigned a conventional telephone number in the central office and is given the same treatment as a land telephone. These are FM systems that use the 152- to 162-MHz VHF or 450- to 460-MHz UHF bands. Until recently, FCC allocations for all mobile radio services (not just mobile telephone) was only 40 MHz, falling in the 25- to 50-, 151- to 174-, and 450- to 470-MHz bands. To relieve the crowding that has developed, the FCC in 1970 made available to public and private land-mobile users an additional 115 MHz between 806 and 947 MHz. Now that active amplification devices are becoming available at reasonably economical prices for 1-GHz operation, expansion of mobile radio service into this new higher band has begun.

A pictoral block diagram of a typical mobile telephone service (MTS) is shown in Fig. 7-13. It is representative of the Harris Corporation, RF Communications Division, 1525IMTS series. The IMTS designator stands for *improved* mobile telephone service, since this system allows for complete direct dialing without any operator assistance. The major elements of the IMTS system shown in Fig. 7-13 include:

> *Terminal unit:* This unit performs the necessary control, signaling and switching functions to interface the local telephone exchange with the radio base station equipment. In addition to its connection to the dial office, the control terminal may also have a trunk connection to a switchboard for operator assistance on certain types of calls.
>
> *Radio base station:* This installation includes the transmitter, receiver, duplexer, and antenna units, and any necessary control equipment for connection with the terminal unit over wire line or carrier facilities. A duplexer is a system whereby transmission and reception may occur concurrently by using two different frequency carriers for each signal. This enables conversation without the "push-to-talk" requirement. Depending on the terrain, antenna height, and required operating range, the required output of the transmitter unit may vary anywhere from 20 to 250 W. When several channels are to be utilized, it may become necessary to add a more complex duplexer/circulator/combiner network to reduce the antenna requirements and at the

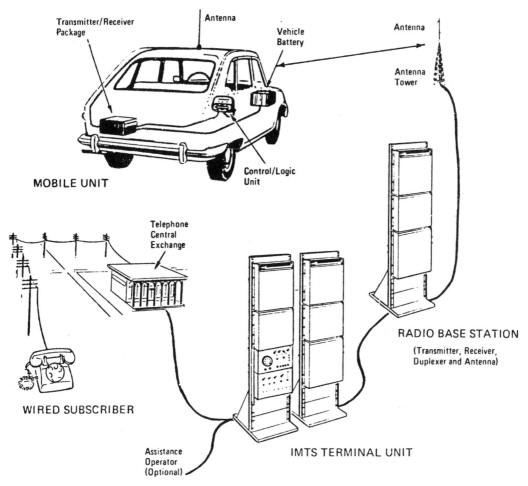

Fig. 7-13. IMTS block diagram (Courtesy of Harris Corp.)

same time protect the system from the possiblility of intermodulation interference. In some cases, it may even be desirable to locate the receiver units at a separate site from their associated transmitters.

Mobile radiotelephones: These units each include a small integrated control/logic unit, which can be located within convenient reach of the vehicle driver, and a compact transmitter/receiver package (including duplexer), which can be located at any out-of-the-way place (e.g., under a seat or in the vehicle luggage compartment). The mobile installation also includes an antenna. The transmitter output power and antenna type are selected to assure adequate communications throughout the required coverage area.

(b)

(a)

(c)

Fig. 7-14. Harris 1525 two-way radiotelephone. (Courtesy of Harris Corp.)

The IMTS system components are pictured in Fig. 7-14. The top photo Fig. 7-14(a) is the transmitter/receiver package. If it is to be located remotely (i.e., in the trunk), it is used in conjunction with the transmitter/receiver control shown in Fig. 7-14(b). The IMTS handset, dial, and control unit is shown in Fig. 7-14(c).

Principles of Operation

The most basic feature of the IMTS system that gives it the flexibility to offer all the rest of its advanced features is the automatic selection and marking of a radio channel for each call. Whenever there are channels idle and available for traffic, the terminal unit selects one and activates the associated base station transmitter modulated by an "idle marking" tone. All idle mobile units scan over their available channels until the idle tone is detected and then lock onto the marked channel. The next call in either direction is then established on this channel with the participating mobile unit remaining locked to it. In the process of completing the call, the terminal unit moves the idle marker to another free channel, again locking idle mobile units to a common channel in readiness for the next call. This feature eliminates the necessity for the mobile users to monitor the channels to find a free one or to maintain watch on a separate calling channel.

If all channels are busy, the mobile unit will illuminate its "busy lamp" when the telephone handset is removed from its cradle. In addition, the RF Communications IMTS unit offers the unique feature of providing the conventional "engaged" signal in the handset ear piece when the "all-channels-busy" situation exists. A fixed subscriber placing a call to a mobile unit when all channels are busy may receive the "all-trunks-busy" tone indication or an optional voice announcement.

To maintain system order, each mobile unit is assigned a unique seven-digit identification and selective calling number. Mobile units will respond only upon receipt of their assigned number. On mobile originated calls, the unit transmits this same number back to the terminal unit as identification so that the call will be routed to the proper line circuit for completion. When the full capacity of the seven identification digits is not required, the number of digits can be reduced accordingly. This can result in a saving in time required for selective calling and identification.

7-6 COMMUNICATION TRANSCEIVER

Communication transceivers span the range from the $50 CB units to the ultra-sophisticated unit pictured in Fig. 7-15, which sells for over $10,000. It is the Harris Corp. RF-280 Transceiver built mainly for the military but also available to civilian users. The specifications are provided in Fig. 7-16.

Fig. 7-15. Harris corp. RF-280 transceiver. (Courtesy of Harris Corp.)

The RF-280 Transceiver is a high-quality FM-SSB transceiver designed and built for compatibility with current military radio communication systems and equipment specifications. The RF-280 provides to the operator all standard modes of tactical two-way radio communications—USB, LSB, FM, AM, and CW—throughout the frequency range of 1.5 to 80.0 MHz.

When used with the automatic antenna couplers, RF-281 (HF) and RF-288 (VHF), the RF-280 provides full automatic antenna coupler tuning and a one-step, one-control transceiver tuning. With this system, the operator can change frequencies, tune, and be transmitting with full power in less than 20 seconds.

When transmitting, a fully protected all-solid-state power amplifier delivers 100 W of RF power in the HF range (1.5 to 30 MHz) to the antenna system. Output power is reduced above 30 MHz to 50 W. The power amplifier protection circuits prevent damage to the solid-state circuitry when an overload condition occurs, such as a shorted or open antenna system.

To reduce to a minimum the possibility of operator error and special training, all complex tuning procedures have been eliminated by the use of automatic circuits, in conjunction with the power amplifier protection circuitry. Initiation and resetting are provided to the operator by front-panel indicators and controls.

SPECIFICATIONS

GENERAL

Frequency Range	1.5 to 80 MHz.
Channel Spacing	785,000 synthesized channels with 100 Hz spacing and $\pm$5kHz continuous VFO tuning.
Frequency Stability	$\pm$1 part in 10^6 (TCXO — Temperature Controlled Crystal Oscillator). Higher stability, optional.
Modes of Operation	AM, LSB, USB, CW, FM and FSK.
Power input	115/230 volts $\pm$ 10%, 50/60 Hz. Plug-in DC power modules can be installed in the unit to provide either 12 or 24 VDC operation in addition to AC operation.
Load Impedance	50 ohms, nominal.
Size	7.65H x 17.22W x 18.0D inches (19.4H x 43.7W x 45.7D cm) (less mounting feet).
Weight	69 pounds (31.3 kg).
Temperature Range	Per Mil-E-16400 Rev F Class II: -28 to +65°C USB, LSB, AM, FM and CW. -28 to +50° FSK Continuous.
Splashproof	Per Mil-Std-108E.
Shock/Vibration (with Shock Mount)	Per Mil-Std-202D Method 201A and Mil-Std-167.
Hammerblow	Per Mil-Std-202D Method 207A and Mil-Std-901C.
Humidity	Per Mil-Std-810B Method 507.

TRANSMITTER

Output Power	1.5-30 MHz 100W PEP and average. 30-80 MHz 50W PEP and average.
Overload Protection	P.A. fully protected from mismatch including an open or shorted antenna.
Audio input	Either carbon or dynamic microphone, 600 ohm input also provided.
Carrier Suppression	50 dB minimum.
Undesired Sideband Suppression	50 dB minimum @ 1kHz.
Harmonic Suppression	40 dB nominal without coupler.
Intermodulation Distortion	-30 dB nominal.
FM Deviation	$\pm$8-10 kHz.

RECEIVER

Sensitivity	SSB: 0.5 μV for 10 dB S + N/N. FM: 0.5 μV for 10 dB SINAD.
Audio Output	Internal speaker: 3 watts with less than 5% distortion. 600 ohm output: 0 dBm minimum with 3 μV input signal.
Selectivity	SSB and FM: 300-3100 Hz at 6 dB. CW: 350 Hz (1 kHz center frequency) (Utilizes crystal filters).
Image and I.F. Rejection	70 db.
AGC	Attack Time: 10 ms nominal. Release Time: SSB-1 sec. nominal. FM—100 ms nominal. Threshold: Internally adjustable 3 μV to 12 μV.
Squelch	Selectable—(1) Variable Threshold (SSB, AM, FM, CW). (2) 150 Hz Tone (FM).
Noise Suppression	Variable threshold noise blanker.
External Spurious Responses	-80 dB.
Internal Spurious Response	99.5% below 0.5 μV equivalent signal at antenna input.
AM Rejection (in FM mode)	Better than 40 dB.
Overload Protection	Withstands 100W input signal at antenna terminals without damage.
Antenna Radiation	Less than 100 picowatts into 50 ohms.

Specifications subject to change without notice

Fig. 7-16. RF-280 specifications. (Courtesy of Harris Corp.)

223

A total of 785,000 separate frequency channels are provided by the use of a solid-state synthesizer in conjunction with a TCXO (temperature-compensated crystal oscillator) frequency standard. Frequency selection is fast, accurate, and is selectable down to the 100-Hz digit of any desired operating frequency by six front-panel controls. Additionally, a VFO control with a range of 10 kHz is a standard feature on the RF-280.

Constructed on a single chassis of lightweight aluminum, the RF-280 weighs less than 70 lb (32 kg). An easy removable wrapover top cover protects the front-panel controls from damage and completes the dust and splashproof integrity of the unit. With the cover removed, all plug-in modules are accessible for easy and fast replacement.

As a two-way voice and CW transceiver, the RF-280 is fully compatible with current HF and VHF military tactical radios. With its extended frequency range, up to 80 MHz, it is ideally suited for those operations where communication requirements can quickly change, such as from a local line-of-sight VHF network to a long-distance HF link with a remote site or distant base station. The all-in-one unit HF and VHF feature of the RF-280 also makes it a valuable alternative to two separate radio sets. This is especially important when there are space-weight restrictions at the installations site.

The capability of operating on both ac and dc power allows the RF-280 to be used for remote temporary sites on dc power only or as a fixed base station on ac power with back-up dc emergency power. Figure (7-17) illustrates the receive signal flow with a heavy solid line and the transmit signal flow with a heavy broken line. The injection signals for frequency conversion, illustrated by a light solid line, are generated in the synthesizer module. The first and second LO signals are synthesized, controlled by the frequency MHz and KHz controls and phase locked to the 5 MHz frequency standard (TCXO). The third LO signal of 500 KHz is divided down directly from the frequency standard signal. The necessary switching of the first and second LO signals to the mixers in the signal processor module is carried out by the LO switch board.

Receive Signal Flow

The received signal from the antenna is filtered and amplified in the preselector module shown in Fig. (7-17) and converted to the high intermediate (up-conversion) frequency (IF) in the mixer board. The high IF is then filtered in the helical filter board and converted to the low IF in the mixer board. After being filtered in the IF filter board, the low IF is amplified and converted to audio (detected) in the receive board. The speaker driver board then amplifies the audio to drive the front panel speaker.

Tuneable filters in the preselector module reject all but the desired received signal, which is amplified by the RF amplifier and filtered again before being switched to the signal processor module. Preselector module filters are tuned by the front panel TUNE control.

The tuned and amplified RF signal from the preselector module is mixed with the first LO signal in the mixer board to produce one of three high IF signals (156 MHz, 126 MHz or 96 MHz), depending on operating frequency, as shown in Fig. 7-18.

The high IF is filtered in the helical filter board and mixed with the second LO signal in the mixer board to produce the low IF signal of 9 MHz, which is amplified in the IF filter board. The low IF is then filtered and coupled to the receive board where it is amplified and detected. One detector provides the automatic gain control (AGC), set by the RECEIVE RF GAIN, which controls attenuation of the RF signal in the preselector module. The three information detectors (FM, AM, and SSB/CW) extract the audio signal, which is amplified and coupled through a volume control (RECEIVE AUDIO control) to the speaker driver board. The 9-MHz injection required for SSB and CW detection is provided by the transmit board. Audio from the volume control is amplified by the speaker driver board to drive the front-panel speaker.

Transmit Signal Flow

The modulated RF transmit signal from the transmit board shown in Fig. 7-17 is filtered and amplified in the IF filter board. This low-IF signal is then converted to the high IF in the mixer board and filtered by the helical filter board. The high IF is then converted to the operating frequency in the mixer board and coupled to the preselector module, where it is filtered and amplified. The P.A. module further amplifies this RF and directs it to the P.A. filter module for final filtering before it is switched to the antenna.

The third LO signal of 500 kHz is multiplied by 18 in the transmit board to provide the low IF of 9 MHz (transmit signal). The transmit signal is modulated by amplified audio from the microphone and coupled to the IF filter board. The 9-MHz transmit signal is filtered here by a bandpass filter and amplified by a two-stage amplifier. The ALC (automatic level control) signal from the P.A. module controls the gain of the first stage and the ACC (automatic carrier control) signal controls the amount of AM carrier inserted into the second stage. The IF filter board output is converted to the high IF in the mixer board and filtered by the helical filter board before final conversion to the operating frequency (Fig. 7-18) in the mixer board. Tuneable filters in the preselector module reject all frequencies but the desired operating frequency, which is amplified and switched to the P.A. module for further amplification.

The RF power output of the P.A. module final amplifier is filtered by the P.A. filter module to remove harmonics before it is coupled through the VSWR bridge to the antenna. The ALC and ACC control signals are developed from the RF output signal at the VSWR bridge and routed back to the IF filter board.

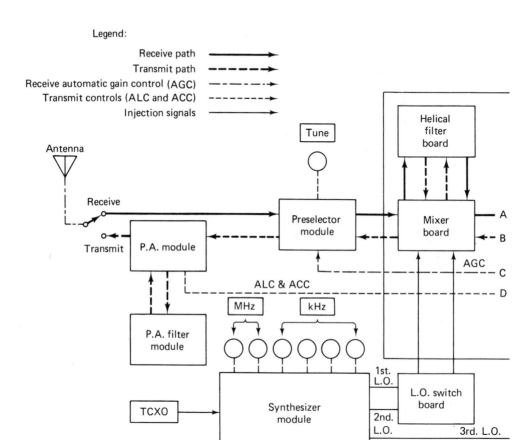

Fig. 7-17. RF-280 signal flow. (Courtesy of Harris Corp.)

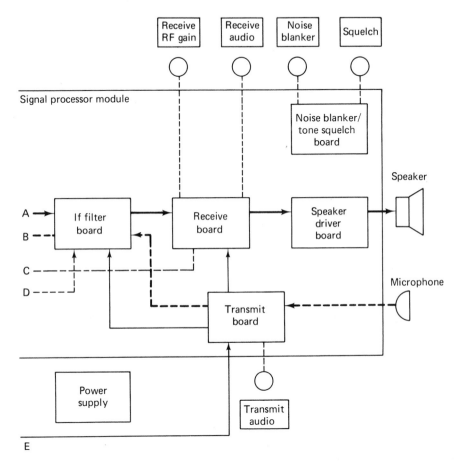

Fig. 7-17. (Cont.).

Operating Freq. MHz	1st L.O. MHz Variable	High IF MHz	2nd L.O. MHz	Low IF MHz
1.5 to 29.9999	157.5 to 185.9999	156	165	9.0
30.0 to 59.9999	156.0 to 185.9999	126	135	9.0
60.0 to 79.9999	156.0 to 185.9999	96	105	9.0

Fig. 7-18. RF-280 internal frequencies. (Courtesy of Harris Corp.)

7-7 COMMUNICATIONS TRANSCEIVER ON A CHIP

A block diagram for a monolithic LIC communications transceiver is provided in Fig. 7-19. The only "outboard" components shown are the two required crystals, one for transmit and one for receive. If multichannel operation is desired, Lithic Systems' LP2700 transceiver chip can easily be driven by a frequency synthesizer instead.

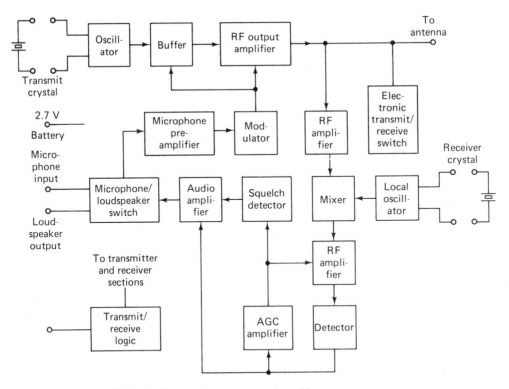

Fig. 7-19. Block diagram for microtransceiver chip.

LITHIC LP2700 MICROTRANSCEIVER

Lithic Systems, Inc.

10101 Bubb Road, P.O. Box 869, Cupertino, California 95014 (408) 257-2004

FEATURES

*Complete transceiver on single bipolar monolithic chip—over 100 transistors.
*100 mW AM input in 27 MHz Citizen's Band; useful to 130 MHz
*Sensitive, single conversion, crystal controlled superhet receiver
*Uses single 2.7 volt supply—can be integrated into wristwatch or other small package configuration.
*End of battery life operation to 2.1 volts.
*Electronic Transmit/Receive switching—no multi-pole switch needed to transfer single antenna, mike/speaker combination—one contact to ground.
*Squelch and volume control provisions.
*Excellent receiver AGC range for large range of input signal strengths.
*Requires minimal external tuned, bypass and control components.
*Drives Speaker/Mike directly—no transformer or large capacitor needed.
*Sensitive mike preamplifier drives controlled-current modulator.
*24 pin DIP package—also available in chip form, for hybrid use.

PERFORMANCE

Parameter (27 MHz, 2.7 volts)	Typical Value	Units
TRANSMITTER:		
Power Input at 27 MHz	100	mW
Power Output	70	mW
Current Drain (Push to Talk)	48	mA
Microphone sensitivity for 100% modulation	1	mV
Frequency stability and spectral purity	(Determined by crystal and external tuned elements)	
RECEIVER:		
Sensitivity for 10 dB (S + N)/N	1	uV
AGC Range for 6 dB audio droop	80	dB
Audio Output Power at 5% max THD into 4 ohm speaker	70	mW
Selectivity (Determined by external ceramic filter)	6	kHz
Current Drain (Squelched)	1.5	mA
Current Drain (Max audio power out, peaks)	27	mA
Audio Distortion (80% modulation, 1 mV in, 1 KHz audio)	2	%
GENERAL		
Battery Voltage Range	2.1 to 3.2	V

Transmit and receive crystals—3rd. overtone CB type—can be driven by
external synthesizer chip with 455 KHz IF offset.

APPROXIMATE EXTERNAL COMPONENTS REQUIRED: 4 LC's, 10 bypass caps., 2 pots., spkr/mike, two battery cells, antenna, 2 crystals per channel.

Fig. 7-20. Microtransceiver features and performance. (Courtesy of Lithic Systems, Inc.)

The LP2700 contains separate transmitter and receiver circuitry as well as the transmit/receive logic circuits. It is operable at frequencies up to 130 MHz. It is an AM system with potential application as a "wrist-worn" CB transceiver, alarm systems, cordless telephones, and telemetry equipment.

Producing a practical wrist transceiver will take more than just dropping the chip (70 by 100 mils when not encapsulated into the dip package) and several tuned circuit elements into a watch case. The input/output transducers—antenna and a microphone/speaker—present an engineering challenge. A possibility for the antenna is to embed an appropriate wire in a plastic wrist band, thus forming a loop-type antenna. The chip is designed to operate off of a 2.7-V battery "button" as is available for digital wristwatches.

Some interesting features and performance characteristics for the LP2700 microtransceiver are provided in Fig. 7-20. It has electronic transmit/receive switching; a single contact to ground transfers the antenna and mike/speaker combination.

An extremely wide AGC range (80 dB) is indicated in Fig. 7-20. An excellent sensitivity for AM of 1 μV is also specified.

8

DIGITAL COMMUNICATIONS

Digital communications refers to a number of different concepts. In this chapter, the following digitally related subjects will be presented:

1. Coding.
2. Radio telegraphy.
3. Pulse modulation.
4. Pulse-code modulation.
5. Digital data.
6. Telemetry.

8-1 CODING

Some of the earliest forms of electrical communication used coding to send messages rather than direct transmission of *voice*. The telegraph demonstration by Samuel Morse in 1843 is an example. It is ironic that digital communications (as telegraphy is) now promises to help ease the problems of overcrowded voice transmission facilities. The future will certainly bring more and more coded speech, transmitted in digital format because of the following advantages:

1. Less sensitive to noise.
2. Less crosstalk (cochannel interference).
3. Lower distortion levels.
4. Faded signals are more easily recreated.

What at first seemed a barrier to digital transmission—that is the need to encode (convert) an analog signal into binary form—is proving to be an advantage. Coding now allows speech to be compressed to its minimum essential content and therefore permits the greatest possible efficiency in transmission. In reality, the previous discussion applies to the coding methods discussed in Sec. (8-4). The more elementary forms to be presented now, however, are still widely used and merit your close attention.

Coding may be defined as the process of transforming messages or signals in accordance with a definite set of rules. There are many different codes available for use but one thing they universally share is the use of two levels. We can then refer to this as a binary system. In such a system, the next signal will either be "high" or "low" and should have an equal (50:50) chance of being one or the other. A *bit* is a unit of information required to allow proper selection of one out of two equally probable events. For example, assuming that a result of heads or tails when flipping a coin are equally probable, let a "high" condition represent heads and a "low" condition represent tails. It is required to have one bit of information to predict the result of the coin toss. If that one bit of information is a "high" condition (usually termed the 1 level), then we can correctly predict that heads came up.

The "high" and "low" conditions are referred to as 1 or 0 or "mark" and "space," respectively. With respect to an electrical signal used to represent these conditions, the relationship is shown in Fig. 8-1. In this figure, 7 bits of informa-

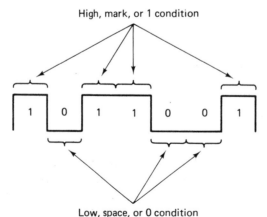

Fig. 8-1. Binary information.

tion are provided with the following sequence: 1 0 1 1 0 0 1. With 7 bits available, it is possible to select the correct result out of 128 equiprobable events, since the 7 bits can be arranged in 128 different configurations and each one may be allowed to represent one of 128 equiprobable events in a code. Since two raised to the seventh power equals 128, we can infer the following relationship:

$$B = \log_2 N \tag{8-1}$$

where B = number of bits of information required to predict one out of N equiprobable events

If it were required to use a binary code to represent the 26 letters in the alphabet, Eq. (8-1) shows that

$$B = \log_2 26$$
$$= 4.7$$

Since the number of bits required must always be an integer, 5 bits are necessary. Since 5 bits means a choice out of 32 different events ($2^5 = 32$) is possible, not all the capability of a 5-bit system is utilized. The efficiency is

$$\mu = \frac{4.7}{5} \times 100\% = 94\%$$

It is not absolutely necessary to use a binary system for coding but it is easy to show why it is used almost exclusively. A decimal system or ten different levels is sometimes utilized. In it, the *dit* (decimal digit) is the basic unit of information. If it were used to represent our alphabet, the number of required dits would be $\log_{10} 26 = 1.415$ dits. Thus, two dits would be required to represent the 26 different letters which means the efficiency of the decimal system in this case would be

$$\mu = \frac{1.415}{2} \times 100\% = 71\%$$

It is a fact that binary coding is more efficient than others except in isolated instances.

EXAMPLE 8-1

Determine the number of bits required for a binary code to represent 110 different possibilities and compare its efficiency with a decimal system to accomplish the same goal.

Solution:

In a binary system

$$B = \log_2 N \tag{8-1}$$
$$= \log_2 110$$
$$= 6.78$$

(The solution to Eq (8-1) can be accomplished by taking the natural log of 110 divided by the natural log of 2: that is, ln 110/ln 2 = 6.78, and therefore, $2^{6.78} = 110$.) Thus, 7 bits are required and the efficiency is

$$\mu = \frac{6.78}{7} \times 100\% = 97\%$$

In a decimal system, the number of dits required is $\log_{10} 110 = 2.04$, or a total of 3 dits. Efficiency is $\mu = 2.04/3 \times 100\% = 68\%$.

Code Noise Immunity

We have seen that binary coding systems are generally more efficient than others. Perhaps an even greater advantage is their greater noise immunity. Consider a binary system where O V represents 0 and 9 V represents 1. In that system it would take a noise level of roughly 5 V or more to cause an error in output intelligibility. In a decimally coded system of the same total output power, 0 is represented by 0 V, 1 by 1 V, 2 by 2 V, etc., up to 9 by 9 V. Thus, the 10 discrete levels are 1 V apart and a 0.5-V noise level can impair intelligibility. If the output levels in this decimal system were 0, 10 V, 20 V, . . . , 90 V, then the binary and decimal system would have comparable noise immunities. Since power is proportional to the square of voltage, the decimal system requires 10^2, or 100, times the power of the binary system to offer the same noise immunity. By now, in your study of communications, you may have come to the correct conclusion that noise is the single most important consideration. The reasoning just concluded regarding efficiency and noise immunity is an elementary example of the work of an information theory specialist. Recall from Chapter 1 that *information theory* is the branch of learning concerned with optimization of transmitted information.

Common Codes

The alphabet has 26 letters, and there is an almost equal number of commonly used symbols and numbers. The 5-bit Baudot code shown in Fig. 8-2 is capable of handling all these possibilities. A 5-bit code can have only 2^5, or 32, bits of information, but actually provides 26×2 bits by transmitting a 11111 to indicate that all following items are "letters" until a 11011 transmission occurs, indicating "figures."

The ARQ code shown in Fig. 8-2 is an adaptation of the Baudot code, which uses 7 bits rather than 5 bits. ARQ is a telegraphic code that means "automatic request for repetition." A 7-bit code allows 2^7, or 128, bits of information, of which only 32 are used. This may seem unnecessarily inefficient—why transmit 7 bits when 5 bits are adequate? Notice that the 32 combinations used in the ARQ code all have three ones and four zeros. There are actually 35 combinations where this is possible. The receiver of ARQ code can detect an error if he detects a mark-to-space ratio other than 3:4, at which time he sends out the ARQ signal—a request for retransmission. The extra time required to transmit ARQ code versus Baudot

(a)

A	1 000 001
B	1 000 010
C	1 000 011
D	1 000 100
E	1 000 101
F	1 000 110
G	1 000 111
H	1 001 000
I	1 001 001
J	1 001 010
K	1 001 011
L	1 001 100
M	1 001 101
N	1 001 110
O	1 001 111
P	1 010 000
Q	1 010 001
R	1 010 010
S	1 010 011
T	1 010 100
U	1 010 101
V	1 010 110
W	1 010 111
X	1 011 000
Y	1 011 001
Z	1 011 010
a	1 100 001
b	1 100 010
c	1 100 011
d	1 100 100
e	1 100 101
f	1 100 110
g	1 100 111
h	1 101 000
i	1 101 001
j	1 101 010
k	1 101 011
l	1 101 100
m	1 101 101
n	1 101 110
o	1 101 111
p	1 110 000
q	1 110 001
r	1 110 010
s	1 110 011
t	1 110 100
u	1 110 101
v	1 110 110
w	1 110 111
x	1 111 000
y	1 111 001
z	1 111 010
0	0 110 000
1	0 110 001
2	0 110 010
3	0 110 011
4	0 110 100
5	0 110 101
6	0 110 110
7	0 110 111
8	0 111 000
9	0 111 001
SP	0 100 000
!	0 100 001
"	0 100 010
#	0 100 011
$	0 100 100
%	0 100 101
&	0 100 110
'	0 100 111
(	0 101 000
)	0 101 001
*	0 101 010
+	0 101 011
,	0 101 100
-	0 101 101
.	0 101 110
/	0 101 111

(b)

A	—	00 011
B	?	11 001
C	:	01 110
D	$	01 001
E	3	00 001
F	!	01 101
G	8	11 010
H	#	10 100
I	8	00 110
J	'	01 011
K	(	01 111
L	)	10 010
M	.	11 100
N	,	01 100
O	9	11 000
P	0	10 110
Q	1	10 111
R	4	01 010
S	Bell	00 101
T	5	10 000
U	7	00 111
V	;	11 110
W	2	10 011
X	/	11 101
Y	6	10 101
Z	"	10 001

(NO LOWER CASE)
"LETTERS" 11 111 1F
"FIGURES" 11 011 1B

(c)

Figures	Letters							
—	A	0	0	1	1	0	1	0
?	B	0	0	1	1	0	0	1
:	C	1	0	0	1	1	0	0
Who are you?	D	0	0	1	1	1	0	0
3	E	0	1	1	1	0	0	0
%	F	0	0	1	0	0	1	1
@	G	1	1	0	0	0	0	1
£	H	1	0	1	0	0	1	0
8	I	1	1	1	0	0	0	0
Bell	J	0	1	0	0	0	1	1
(	K	0	0	0	1	0	1	1
)	L	1	1	0	0	0	1	0
.	M	1	0	1	0	0	0	1
,	N	1	0	1	0	1	0	0
9	O	1	0	0	0	1	1	0
φ	P	1	0	0	1	0	1	0
1	Q	0	0	0	1	1	0	1
4	R	1	1	0	0	1	0	0
'	S	0	1	0	1	0	1	0
5	T	1	0	0	0	1	0	1
7	U	0	1	1	0	0	1	0
=	V	1	0	0	1	0	0	1
2	W	0	1	0	0	1	0	1
/	X	0	0	1	0	1	1	0
6	Y	0	0	1	0	1	0	1
+	Z	0	1	1	0	0	0	1
Carriage return		1	0	0	0	0	1	1
Line feed		1	0	1	1	0	0	0
Figures shift		0	1	0	0	1	1	0
Letters shift		0	0	0	1	1	1	0
Space		1	1	0	1	0	0	0
Unperforated tape		0	0	0	0	1	1	1

Fig. 8-2. Common codes: (a) 7-bit ASC 11 code; (b) 5-bit Baudot; (c) ARQ code.

is justified under noisy conditions. This is an elementary example of redundancy. *Redundancy* may be defined as the fraction of the gross information content of a message that can be eliminated without loss of essential information. A *redundant code* is a code using more signal elements than necessary to represent the intrinsic information. Redundancy is introduced into an information system to either allow intelligibility when part of the signal is mutilated by noise or to alert the receiver that mutilation has occurred and to be wary of the received signal's reliability.

The previous codes do not have provision for upper- *and* lowercase letters. The 7-bit ASCII code shown in Fig. 8-2 can be used when this may be a requirement.

Figure 8-3(a) shows an example of the Baudot code to transmit "YANKEES 4 REDSOX 3." Be sure to work out the code in Fig. 8-3(b) on your own; it is the only "x-rated" part of this book I was allowed to include.

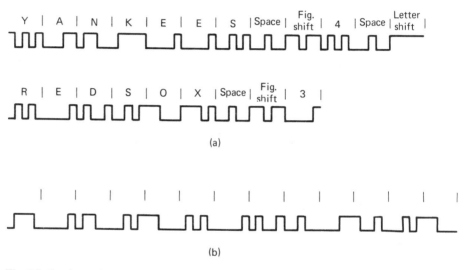

(a)

(b)

Fig. 8-3. Baudot code examples.

Morse Code

The familiar Morse code is not a true binary code, in that it not only includes marks and spaces but also differentiates between the duration of these conditions. The Morse code is, therefore, not well suited to today's automated telegraphic equipment but is still widely used in amateur radio-telegraphic communication. The machines used are very good at differentiating between a "high" or "low" but to also require them to make differentiations between various widths of pulses is an extreme complicating factor at the receiver and prone to error. However, a human skilled at code reception can provide highly accurate decoding. The International Morse Code is shown in Fig. 8-4. It consists of dots (short mark), dashes (long

Digital Communications

```
A  ·—            N  —·            1  ·————
B  —···          O  ———           2  ··———
C  —·—·          P  ·——·          3  ···——
D  —··           Q  ——·—          4  ····—
E  ·             R  ·—·           5  ·····
F  ··—·          S  ···           6  —····
G  ——·           T  —             7  ——···
H  ····          U  ··—           8  ———··
I  ··            V  ···—          9  ————·
J  ·———          W  ·——           0  —————
K  —·—           X  —··—
L  ·—··          Y  —·——
M  ——            Z  ——··
```

```
     (period)                    ·—·—·—
,    (comma)                     ——··——
?    (question mark)(I̅M̅I̅)       ··——··
/    (fraction bar)              —··—·
:    (colon)                     ———···
;    (semicolon)                 —·—·—·
(    (parenthesis)               —·——·—
)    (parenthesis)               —·——·
'    (apostrophe)                ·————·
-    (hyphen or dash)            —····—
$    (dollar sign)               ···—··—
"    (quotation marks)           ·—··—·
```

Fig. 8-4. International Morse code.

mark), and spaces. A "dot" is made by pressing the telegraph key down and allowing it to spring back rapidly. The length of a dot is one basic time unit. The "dash" is made by holding the key down (keying) for three basic time units. The spacing between dots and dashes in one letter is one basic time unit and between letters is three units. The spacing between words is seven units.

8-2 CODE TRANSMISSION

The most elementary form of transmitting "highs" and "lows" is to simply key a transmitter's carrier on and off. Figure 8-5(a) shows a dot–dash–dot waveform, while Fig. 8-5(b) shows the resulting transmitter output if the "mark" allows the carrier to be transmitted and "space" cuts off transmission. Thus, the carrier is conveying intelligence by simply turning it on or off according to a prearranged

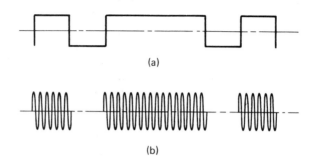

(a)

(b)

Fig. 8-5. CW waveforms.

code. This type of transmission is called *continuous wave* (CW): however, since the wave is periodically interrupted, it might more appropriately be called *interrupted wave*. As a concession to the CW misnomer, it is sometimes called *interrupted continuous wave* (ICW).

Whether the CW shown in Fig. 8-5(b) is created by a hand-operated key, remote controlled relay, or an automatic system such as punched tape, the rapid rise and fall of the carrier presents a problem. The steep sides of the waveform are rich in harmonic content, which means that the channel bandwidth for transmission would have to be extremely wide, or else adjacent channel interference would occur. This is a severe problem in that a major advantage of coded transmission-versus-direct voice transmission is that of the narrow bandwidth channels. The situation is remedied by use of a simple low-pass *LC* filter, as shown in Fig. 8-6. The inductor slows down the rise time of the carrier while the capacitor slows down the decay. This filter is known as a *keying filter* and is also effective in blocking transmission of the RFI (radio-frequency interference) created by arcing of the key contacts.

CW is a form of AM and therefore suffers from noise to a much greater extent than FM systems. The space condition (no carrier) is also troublesome to a receiver, since at that time the receiver's gain is increased by AGC action so as to make received noise very troublesome. Manual receiver gain control helps but not if the received signal is fading between high and low levels, as is often the case. Its simplicity and narrow bandwidth makes it attractive to radio amateurs, but commercial users resort to two other types of code transmission.

Two-Tone Modulation

Two-tone modulation is also a form of AM but in it, the carrier is always transmitted. Instead of simply turning the carrier on and off, the carrier is amplitude-modulated by two different frequencies, representing either mark or space. The two frequencies are usually separated by 170 Hz. An example of such a telegraphy system is provided in Fig. 8-7. When the transmitter is keyed, the carrier is modulated by a 470-Hz signal (mark condition) while it is modulated by a 300-Hz

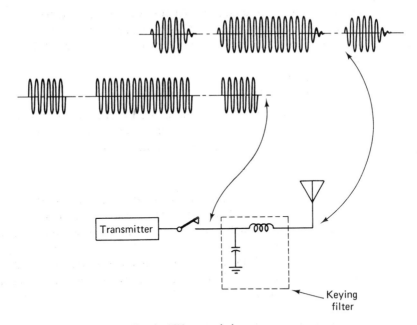

Fig. 8-6. Keying filter effect in CW transmission.

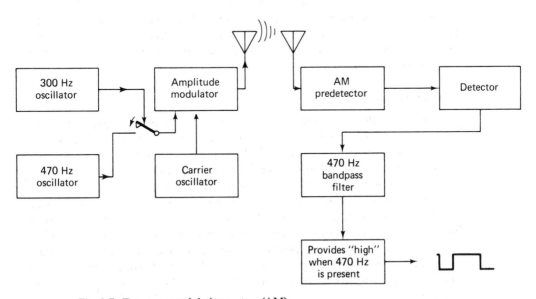

Fig. 8-7. Two-tone modulation system (AM).

239

signal for the space condition. At the receiver, after detection, either 300-Hz or 470-Hz signals are present. A 470-Hz bandpass filter provides an output for the mark condition that makes the output high whenever 470 Hz is present and low otherwise.

EXAMPLE 8-2

The two-tone modulation system shown in Fig. 8-7 operates with a 10-MHz carrier. Determine all possible transmitted frequencies and the required bandwidth for this system.

Solution:

This is an amplitude modulation system and, therefore, when the carrier is modulated by 300 Hz, the output frequencies will be 10 MHz and 10 MHz $\pm$ 300 Hz. Similarly, when modulated by 470 Hz, the output frequencies will be 10 MHz and 10 MHz $\pm$ 470 Hz. Those are all possible outputs for this system. The bandwidth required is, therefore, 470 Hz $\times$ 2 = 940 Hz, which means that a 1-kHz channel would be adequate.

Example 8-2 shows that two-tone modulation systems are very effective with respect to bandwidth utilized. One hundred 1-kHz channels could be sandwiched in the frequency spectrum from 10 MHz to 10.1 MHz! The fact that carrier is always transmitted eliminates the receiver gain control problems previously mentioned and the fact that three different frequencies (carrier, USB, LSB) are always being transmitted is another advantage over CW systems. In CW either one frequency, the carrier, or none is transmitted. Single-frequency transmissions are much more subject to ionospheric fading conditions than are multifrequency transmissions.

Frequency-Shift Keying

As you may have expected, a telegraphy system based on frequency modulation techniques offers even better performance than either of the two previously discussed AM systems. *Frequency-shift keying* (FSK) is a form of frequency modulation in which the modulating wave shifts the output between two predetermined frequencies, usually termed the mark and space frequencies. It may be considered as an FM system in which the carrier frequency is midway between the mark and space frequencies and is modulated by a rectangular wave as shown in Fig. 8-8. The mark condition causes the carrier frequency to increase by 42.5 Hz while the space condition results in a 42.5-Hz downward shift. Thus, the transmitter frequency is constantly changing by 85 Hz as it is keyed. This 85-Hz shift is the standard for narrow-band FSK, while an 850-Hz shift is the standard for wide-band FSK systems.

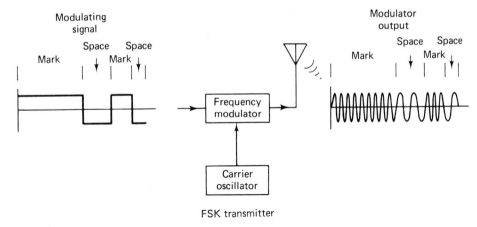

FSK transmitter

Fig. 8-8. FSK transmitter.

EXAMPLE 8-3

Determine the channel bandwidth required for the narrow-band and wide-band FSK systems.

Solution:

The fact that narrow-band FSK shifts a total of 85 Hz does *not* mean that the bandwidth is 85 Hz. While shifting 85 Hz it creates an infinite number of side bands with the extent of significant side bands determined by the modulation index. Thus, the solution of this question requires more information. If this is difficult for you to accept, it would be wise to review the basics of FM in Chapter 5.

In practice, most narrow-band FSK systems utilize a channel of several kilohertz, while wide-band FSK uses 10 to 20 kHz. Because of the narrow bandwidths involved, FSK systems offer only slightly improved noise performance over the AM two-tone modulation scheme. However, the greater numbers of side bands transmitted in FSK allows even better ionospheric fading characteristics than two-tone AM modulation schemes.

FSK Generation

The generation of FSK can be easily accomplished by switching an additional capacitor into the tank circuit of an oscillator when the transmitter is keyed. In narrow-band FSK, it is often possible to get the required frequency shift by shunting the capacitance directly across a crystal, especially if frequency multipliers follow the oscillator, as is usually the case. FSK can also be generated by applying the rectangular wave-modulating signal to a VCO. Such a system is shown in Fig. 8-9. The VCO output is the desired FSK signal, which is then transmitted to an FM

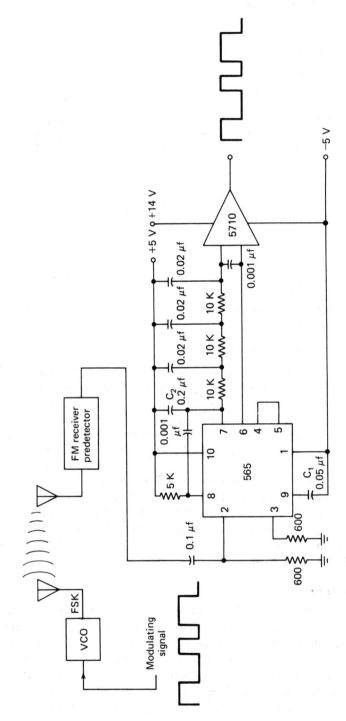

Fig. 8-9. Complete FSK system.

receiver. The receiver is a standard unit up through the IF amps. At that point, a 565 PLL is used for detecting the original modulating signal. As the IF output signal appears at the PLL input, the loop locks to the input frequency and tracks it between the two frequencies with a corresponding dc shift at its output, pin 7. The loop filter capacitor, C_2, is chosen to set the proper overshoot on the output, and the three-stage ladder filter is used to remove the sum-frequency component. The PLL output signal is a rounded-off version of the original binary modulating signal and is therefore applied to the comparator circuit (5710 in Fig. 8-9) to make it logic compatible.

Code Transmission Detail

The most common means of handling telegraphic transmissions is through use of a *teleprinter*. It is a transceiver that not only handles the radio-related functions but has a typewriter-like keyboard. They are commonly referred to as *Teletype* machines, which is a trade name of a major manufacturer of this equipment. When a key is depressed, the machine mechanically creates the appropriate code for the desired letter or symbol, which is subsequently processed for transmission either to an antenna or over standard telephone lines. Another teletype machine receiving the signal decodes it and provides a typewriter-style copy. The transmission medium can be via antennas (radio-teletype) or by standard telephone lines.

The maximum rate at which information can be transmitted is limited by operator typing speed and/or the bandwidth being used. The faster an operator types the shorter must be the duration of the pulses in the transmitted binary code. Of course, shorter pulses mean higher-frequency content in the pulses, since frequency is inversely proportional to pulse width. As a result of this, a measure of telegraph speed is expressed as the reciprocal of time for the shortest transmitted pulse. This time is defined as a *baud*. One baud is the time for the shortest signaling element and speed is expressed in bauds per second.

The standard shortest duration pulse in telegraphy is 22 ms, and its reciprocal, $\frac{1}{22}$ ms—45 bauds per second—is the system's speed. In addition, each letter (coded) is preceded by a 22-ms space and followed by a 33-ms mark. Since the average word has six letters, the average time for each word using a standard 5-bit code is

$$6 \times [22 \text{ ms} + (5 \times 22 \text{ ms}) + 33 \text{ ms}] = 1 \text{ s}$$

This implies a transmission speed of 60 words per minute, since each word, on the average, takes 1 s.

To ensure that a channel is used at its maximum capacity, most teletypes have a perforated tape unit. The operator can then type the message at any convenient speed without transmitting. The message is stored on a perforated tape and then can be transmitted at a convenient time at whatever maximum speed the system has. This provides maximum channel usage efficiency. Figure 8-10 illustrates a commonly used code, the CCITT-2, as it appears on perforated tape. This code is

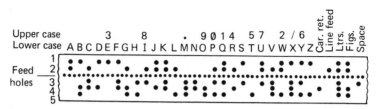

Fig. 8-10. Perforated tape—CC 1TT-2 code.

an adaptation of the Baudot code shown in Fig. 8-2. Each hole (perforation) in the tape represents a mark for that particular bit. The lack of a hole represents a space condition.

8-3 PULSE MODULATION

You have undoubtedly drawn graphs of *continuous* curves many times. To do that, you took data at some finite number of discrete points, plotted each point, and then drew the curve. Drawing the curve may have resulted in a very accurate replica of the desired function, even though you did not look at every possible point. In effect, you took *samples* and guessed where the curve went in between the samples. If the samples had sufficiently close spacing, the result is adequately described. It is possible to apply this line of reasoning to the transmission of an electrical signal: that is, to transmit just the samples and let the receiver reconstruct the total signal with a high degree of accuracy. This is termed *pulse modulation.*

The key distinction between pulse modulation and normal AM or FM is that, in AM or FM, some parameter of the modulated wave varies continuously with the message whereas in pulse modulation some parameter of a sample pulse is varied by each sample value of the message. The pulses are usually of very short duration so that a pulse modulated wave is "off" most of the time. This factor is the main reason for using pulse modulation since it allows:

1. Transmitters to operate on a very low duty cycle as is desirable for certain microwave devices and lasers, or
2. The time intervals between pulses can be filled with samples of other messages.

The latter of the above is the main reason to use pulse modulation in that it conveniently allows a number of different messages to be transmitted on the same channel. This is a form of multiplexing known as *time-division multiplexing* (TDM).

It can be mathematically proven that a signal sampled at twice the rate of its highest significant frequency component can be fully reconstructed at the receiver to a high degree of accuracy. In the case of voice transmission, the standard sampling rate is 8 KHz, it being just slightly more than twice the highest significant frequency component. This implies a pulse rate of 1/8 KHz or 125μs. Since a pulse *duration* of 1 μsec may be adequate, it is easy to see that a number of different messages could be multiplexed (TDM) on the channel or else it would allow a high peak transmitted power with a much lower (1/125) average power. The latter would allow for a very high signal-to-noise ratio or a greater transmission range.

The student should be careful to realize that a price must be paid for system gains obtained by pulse modulation schemes. More important than the greater equipment complexity is the requirement for greater channel (bandwidth) size. If a maximum 3 KHz signal directly amplitude modulates a carrier, a 6 KHz bandwidth is required. If a 1 μsec pulse does the modulating, just allowing its fundamental component of 1/1 μs or 1 MHz to do the modulating means a 2 MHz bandwidth is required in AM! In spite of the large bandwidth required, TDM is still much preferable (if not the only possible way) to using 100 different transmitters, antennas, or transmission lines, and receivers in cases where large numbers of messages must be conveyed simultaneously.

In its strictest sense, pulse modulation is not modulation but rather a *message-processing* technique. The message to be transmitted is sampled by the pulse, and the pulse is subsequently used to either amplitude- or frequency-modulate the carrier. The three basic forms of pulse modulation are illustrated in Fig. 8-11. There are numerous varieties of pulse modulation, and no standard terminology has yet evolved. The three types we shall consider here are usually termed *pulse-amplitude modulation* (PAM), *pulse-duration modulation* (PDM), and *pulse-position modulation* (PPM). For the sake of clarity, the illustration of these modulation schemes has greatly exaggerated the pulse widths. Since the major application of pulse modulation occurs when TDM is to be used, shorter pulse durations, leaving room

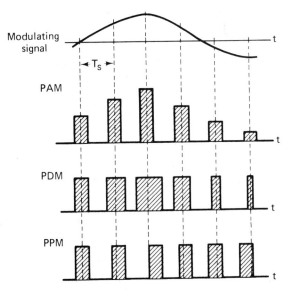

Fig. 8-11. Types of pulse modulation.

for more multiplexed signals, is obviously required. As shown in Fig. 8-11, the pulse parameter that is varied in step with the analog signal is varied in direct step with the signal's value at each sampling interval. Notice that the pulse amplitude in PAM and pulse width in PDM is not zero when the signal is zero. This is done to allow a constant pulse rate and is important in maintaining synchronization in TDM systems.

Pulse-Amplitude Modulation

In pulse-amplitude modulation (PAM), the pulse amplitude is made proportional to the modulating signal's amplitude. This is the simplest pulse modulation to create in that a simple sampling of the modulating signal at a periodic rate can be used to generate the pulses which are subsequently used to modulate a high-frequency carrier. An eight-channel TDM PAM system is illustrated in Fig. 8-12. At the transmitter, the eight signals to be transmitted are periodically sampled. The sampler illustrated is a rotating machine making periodic brush contact with each signal. A similar rotating machine at the receiver is used to distribute the eight separate signals, and it must be synchronized to the transmitter. A mechanical sampling system is suitable for low sampling rates such as are encountered in some telemetry systems but would not be adequate for the 8-kHz rate required for voice transmissions. In that case, an electronic switching system would be incorporated.

At the transmitter, the variable amplitude pulses are used to frequency-modulate a carrier. A rather standard FM receiver recreates the pulses which are then applied to the electromechanical "distributor" going to the eight individual channels. This "distributor" is virtually analogous to the distributor in a car that delivers high voltage to eight spark plugs in a periodic fashion. The pulses applied to each line go into an envelope detector that serves to recreate the original signal. This can be a simple low-pass *RC* filter such as is used following the detection diode in a standard AM receiver.

While PAM finds some use due to its simplicity, PDM and PPM use constant-amplitude pulses and provide superior noise performance. The PDM and PPM systems fall into a general category termed *pulse-time modulation* (PTM) in that their timing, and not amplitude, is the varied parameter.

Pulse-Duration Modulation

Pulse-duration modulation (PDM), a form of PTM is also known as *pulse-width modulation* (PWM) and *pulse-length modulation* (PLM). A simple means of PDM generation is provided in Fig. 8-13 using a 565 PLL. It actually creates PPM at the VCO output (pin 4), but by applying it and the input pulses to an exclusive-or gate, PDM is created also. For the phase-locked loop (PLL) to remain locked, its VCO input (pin 7) must remain constant. The presence of an external modulating signal upsets the equilibrium. This causes the phase detector output to go up or

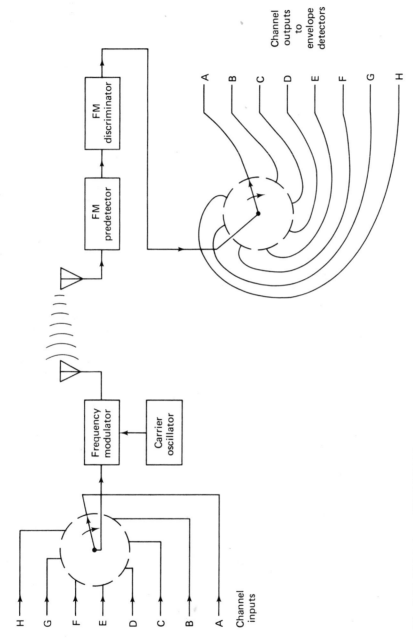

Fig. 8-12. Eight-channel TDM PAM system.

247

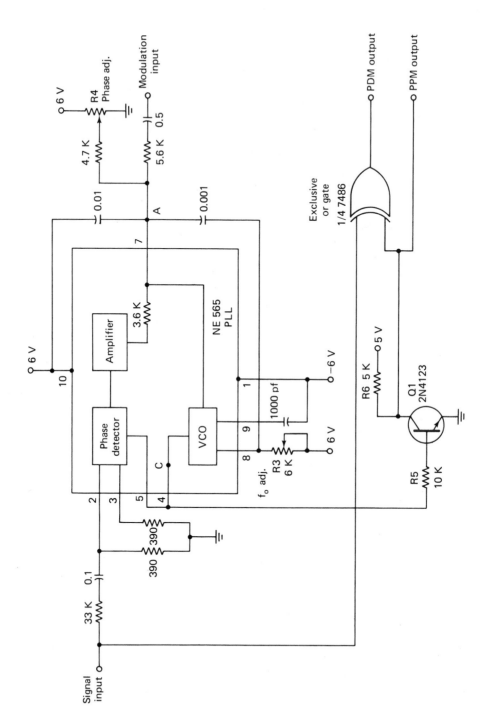

Fig. 8-13. PLL generation of PDM and PPM.

down to maintain the VCO input (control) voltage. However, a change in phase-detector output also means a change in phase difference between the input signal and the VCO signal. Thus, the VCO output has a phase shift proportional to the modulating signal amplitude. This PPM output is amplified by Q_1 in Fig. 8-13 just prior to the output. The exclusive-or circuit provides a "high" output only when just one of its two inputs are "high." Any other input condition produces a "low" output. By comparing the PPM signal and the original pulse input signal as inputs to the exclusive-or circuit, the output is a PDM signal at twice the frequency of the original signal input pulses.

Adjustment of R_3 varies the center frequency of the VCO. The R_4 potentiometer may be adjusted to set-up the quiescent PDM duty cycle. The outputs (PPM or PDM) of this circuit may then be used to modulate a carrier for subsequent amplification and transmission.

Class D Amplifier and PDM Generator

PDM forms the basis for a very efficient form of power amplification. The circuit in Fig. 8-14 is a so-called Class D amplifier since the actual power amplification is provided to the PDM signal and, since it is of constant amplitude, the transistors used can function between cutoff and saturation. This allows for maximum efficiency (in excess of 90%) and is the reason for the increasing popularity of Class D amplifiers as a means of amplifying any analog signal.

The circuit of Fig. 8-14 illustrates another common method for generation of PDM and also illustrates Class D amplification. The Q6 transistor generates a constant current to provide a linear charging rate to capacitor C2. The unijunction transistor, Q5, discharges C2 when its voltage reaches Q5's firing voltage. At this time, C2 starts to charge again. Thus, the signal applied to Q7's base is a linear sawtooth as shown at A in Fig. 8-15. That sawtooth, following amplification by the Q7 emitter follower in Fig. 8-14, is applied to the op amp's inverting input. The modulating signal or signal to be amplified is applied to its non-inverting input which causes the op-amp to act as a comparator. When the sawtooth waveform at A in Fig. 8-15 is less than the modulating signal B, the comparator's output (C) is high. At the instant A becomes greater than B, C goes low. The comparator (op amp) output is, therefore, a PDM signal. It is applied to a push-pull amplifier (Q1, Q2, Q3, Q4) in Fig. 8-14 which is a highly efficient switching amplifier. The output of this power amp is then applied to a low-pass LC circuit (L1, C1) that converts back to the original signal (B) by integrating the PDM signal at C as shown at D in Fig. 8-15.

The output of the op-amp in Fig. 8-14 would be used to modulate a carrier in a communication system while a simple integrating filter would be used at the receiver as the detector to convert from pulses to the original analog modulating signal.

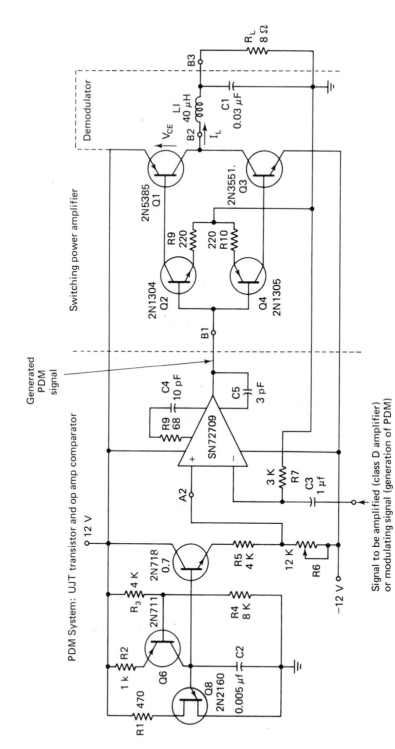

Fig. 8-14. PDM Generator and Class D power amplifier.

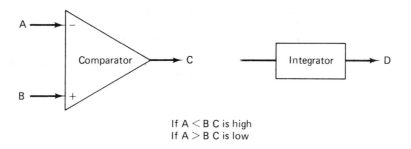

If A < B C is high
If A > B C is low

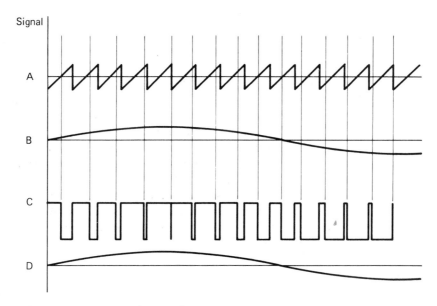

Fig. 8-15. PDM generation waveforms.

Pulse-Position Modulation

Pulse-position modulation (PPM) and PDM are very similar, a fact that is underscored in Fig. 8-16, which shows PPM being generated from PDM. Since PPM has superior noise characteristics, it turns out that the major use for PDM is to generate PPM. By inverting the PDM pulses in Fig. 8-16 and then differentiating them, the positive and negative spikes shown are created. By applying them to a Schmidt trigger sensitive to only positive levels, a constant amplitude and pulse width signal is formed. However, the position of these pulses is variable and now proportional to the original modulating signal and the desired PPM signal has been generated. The information content is *not* contained in either the pulse amplitude or width as in PAM and PDM, which means that the signal now has a greater

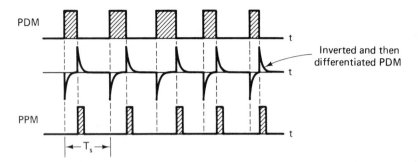

Fig. 8-16. PPM generation.

resistance to any error caused by noise. In addition, when PPM modulation is used to amplitude-modulate a carrier, a power savings results, since the pulse width can be made very small.

At the receiver, the detected PPM pulses are usually converted to PDM first and then converted to the original analog signal by integrating as previously described. Conversion from PPM to PDM can be accomplished by feeding the PPM signal into the base of one transistor in a flip-flop. The other base is fed from synchronizing pulses at the original (transmitter) sampling rate. The period of time that the PPM-fed transistor's collector is low depends on the difference in the two inputs, and it is, therefore, the desired PDM signal.

This detection process illustrates the one disadvantage PPM has compared to PAM and PDM. It requires a pulse generator synchronized from the transmitter. However, its improved noise and transmitted power characteristics make it the most desirable pulse modulation scheme.

8-4 PULSE-CODE MODULATION

Prior to this chapter, the voice transmission schemes studied involved the following process:

1. Voice to electrical signal conversion via a transducer such as a microphone.
2. Continuous modulation of a carrier by the electrical representation of voice.
3. Amplification and subsequent transmission via an antenna.

This system is illustrated in Fig. 8-17(a).

In this chapter, two other methods have already been introduced. The first involved converting each letter of every word into a code and then modulating a carrier with the code. This process, telegraphy, is illustrated in Fig. 8-17(b). The process of pulse modulation discussed in Sec. (8-3) is provided in Fig. 8-17(c). In

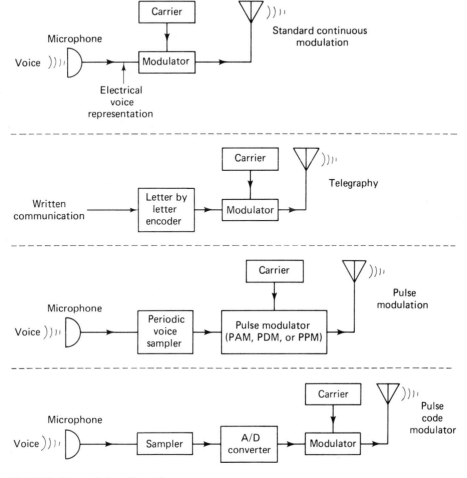

Fig. 8-17. Transmission alternatives.

this case, the electrical voice representation is sampled and the samples are used to vary either a pulse's amplitude, duration, or position. The resulting pulses modulate a carrier, which is then transmitted.

The fourth transmission method [Fig. 8-17(d)] is termed *pulse-code modulation* (PCM). It is generally defined as a modulation process that involves the conversion of a waveform from analog to digital form by means of coding. PCM is a true digital process as compared to the pulse-modulation schemes of Sec. (8-3). In PCM, the electrical representation of voice is converted from analog form to digital form, the binary code. This process of encoding is shown in Fig. 8-18. There are a set of amplitude levels and sampling times. The amplitude levels are termed *quantizing levels*, and 12 such levels are shown. At each sampling interval the analog

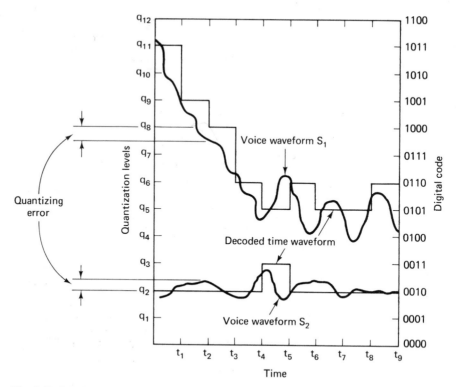

Fig. 8-18. PCM encoding.

amplitude is quantized into the closest available quantization level and the A/D converter puts out a series of pulses representing that level.

For example, at time t_2 in Fig. 8-18, voice waveform, S_1, is closest to level q_8, and thus the coded output at that time is the binary code 1000, which represents 8 in binary code. Note that the quantizing process resulted in an error, which is termed the *quantizing error*. Voice waveform S_2 provides a 0010 code at time t_2 and its quantizing error is also shown in Fig. 8-18. The amount of this error can be minimized by increasing the number of quantizing levels, which, of course, lessens the space between each one. The 4-bit code shown in Fig. 8-18 allows for a maximum of 16 levels, since $2^4 = 16$. The use of a higher-bit code decreases the error at the expense of transmission time and/or bandwidth since, for example, a 5-bit code (32 levels) means transmitting five "high" or "low" pulses instead of four for each sampled point. The sampling rate is also critical and must be at least twice the highest significant frequency, as described for pulse modulation schemes in Sec. (8-3).

While a 4- or 5-bit code may be adequate for voice transmission, it is not adequate for transmission of television signals. Figure 8-19 provides an example of

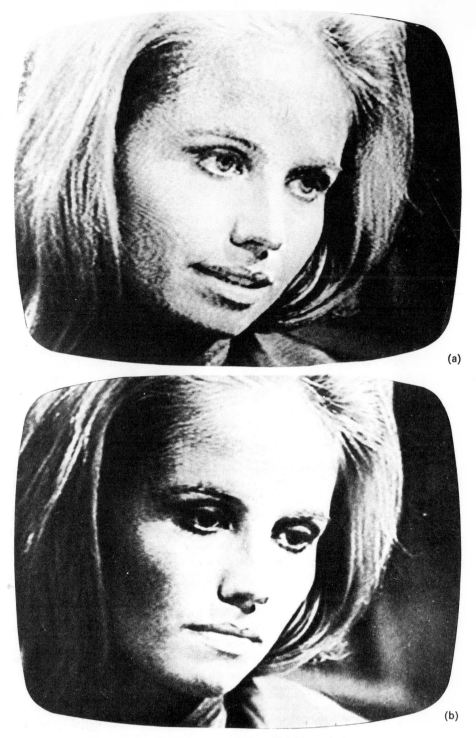

(a)

(b)

Fig. 8-19. PCM TV transmission. (a) 5-bit resolution; (b) 8-bit resolution.

TV pictures for 5-bit and 8-bit (256 levels) PCM transmissions, each with 10-MHz sampling rates. In the first (5-bit) picture, contouring in the forehead and cheek areas is very pronounced. The 8-bit resolution results in an excellent fidelity TV signal that is not discernibly different than a standard continuous modulation transmission.

PCM Advantages

PCM is the most noise-resistant transmission system available. It also lends itself to time-division multiplexing (TDM), and because of its true digital nature, is very adaptable to microprocessor controlled communications equipment, and to the transmission of digital data. For these reasons, PCM is the fastest-growing method of getting information from one place to another. An example of its use is in the distribution of live TV from its point of origination to all parts of the country. To accomplish this, microwave towers at 20- to 30-mile intervals are used to "relay" the TV signals throughout the country. Each tower contains a receiver that receives the signal, reamplifies it, and transmits it to the next tower. The towers are termed *repeaters* and make up for the huge power loss incurred from each transmitter to receiver. Since each reamplification of signal and noise can only aggravate the S/N ratio, it is critical to use the most noise-immune system of modulation possible, PCM. It can be theoretically shown that a S/N ratio of 21 dB or better can be "repeated" indefinitely without any degradation using PCM.

PCM/TDM Repeaters

A PCM/TDM repeater system is shown in Fig. 8-20. It is interesting to notice that the TDM sampler initially creates a PAM signal that is then quantized into an *n*-bit PCM/TDM signal. It is then modulated onto a carrier and transmitted to the first repeater. While only two repeaters are shown, the number can be as large as necessary and is, in fact, in the hundreds for TV signal distribution. At the final destination, the signal is applied to a demodulator which outputs the binary coded pulses to the decoder. The decoder output is a PAM/TDM signal. The time division *de*multiplexer then delivers the three original analog outputs which are applied to low-pass filters that remove unwanted modulation products produced by the demultiplexer. The synchronism link shown in Fig. 8-20 between the multiplexer and demultiplexer is necessary to keep the three distinct signals properly separated.

In actuality, the repeaters are complete demodulators/modulators of the PCM signals. This is the key to infinite repetition without cumulative signal degeneration. With PCM, the signal can be regenerated at each repeater, producing a new signal essentially free of noise. A reference to the pulses shown in Fig. 8-21 serves to illustrate this point. The received PCM signal has been affected by noise but not to the point where it is not recognizable as a 01011 signal. The signal's amplitude, pulse width, and pulse position have been altered by noise and a finite

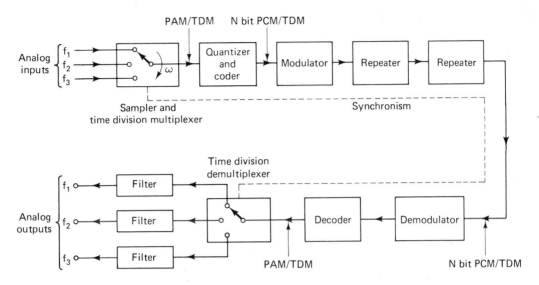

Fig. 8-20. PCM/TDM system.

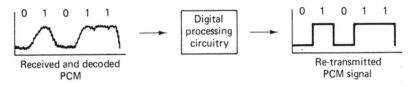

Fig. 8-21. Regeneration of PCM.

bandwidth system that cause pulse rise and fall times to have visible slope. Applying this signal to appropriate digital processing circuitry allows the signal to be "cleaned" up to its original condition for retransmission. This assumes that the noise is not so great as to make it impossible to distinguish between a high or a low on a reliable basis. This process is not possible with standard AM or FM, PAM, PDM, or PPM. The noise errors could not be cleaned up, and cumulative degradation occurs. As an example, with PPM the receiver has no way of knowing that a slightly changed pulse position by noise is, in fact, an error. PCM relies solely on the presence or absence of pulses, not absolute amplitude, duration, or position.

PCM should be given consideration for applications involving TDM, minimum transmitted power, combinations of analog and digital messages, computer control, or many repeater stations. Since long-distance telephone transmission has all these requirements, PCM is the current trend in modern telephony. In routine applications, however, the high cost of hardware for PCM makes it prohibitive compared to analog modulation systems.

8-5 TRANSMISSION OF DIGITAL DATA

There is a gradual transition taking place in types of communication traffic. Formerly, it was mostly analog in nature, but we are now headed to an equal mix between analog and digital signals. This transition has been caused by the increasing use of digital coding (i.e., PCM) for transmission of analog signals and the surging need for computers to talk between themselves and to remote terminals. The links used in these communication systems are normally the standard voice-grade telephone lines, or antenna-to-antenna systems, usually at microwave carrier frequencies.

Designers of the required equipment have been adapting rapidly to this change in traffic mix and tremendous upsurge in volume, with the following changes:

1. The use of computer control systems to find the unused portion of a multiplexed system to maximize use of available channels.
2. The application of digital switching theory to increase channel capacity.
3. Sharing of communication links by voice and data signals.

Microcomputers (microprocessors) are being used more frequently, even in relatively small, portable-type communications equipment. These computers on a chip serve as network controllers, are used for frequency synthesis, diagnostic tools, digital filters, storage devices, signal processors, and can be used to automatically make many of the adjustments normally required of an operator.

Telephone Line Transmission

The largest communications network in the world, by far, is the telephone system. It also turns out to be the most economical link for transmitting digital data when a large volume of data traffic between two locations does not exist. The use of telephone lines to carry digital data presents several problems, however, and therefore requires the use of complex processing circuitry. The problems arise for the following reasons:

1. The bandwidth of individual telephone lines is barely 2100 Hz, they do not pass dc, and they introduce considerable phase distortion and noise.
2. The data to be transmitted are usually parallel data from a number of lines in the computer and therefore cannot be directly transmitted over a two-wire transmission line.

Because of the limited bandwidth, the steep-sided, flat-topped pulses connot be directly transmitted. The pulses must be used to modulate a carrier that the 2100-Hz-bandwidth lines can handle. Either FSK or phase modulation is used for this process. In addition, the parallel data format must be converted to serial form for transmission over two-wire lines. Although these conversions may sound simple, it turns out that the complexities involved have only recently been made economically attractive, through integration (MSI and LSI) of these functions onto one or two chips.

LSI UART

The process for handling digital data for transmission is shown in block diagram form in Fig. 8-22. The computer outputs are accepted by the *universal*

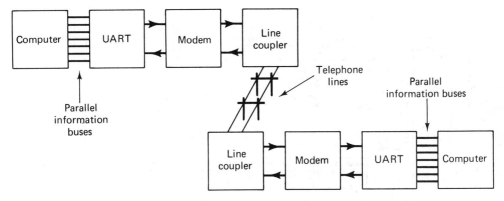

Fig. 8-22. Computer data transmission system.

asynchronous/synchronous receiver/transmitter (UART). The UART, in simplified terms, converts the parallel (asynchronous) computer data into the required serial (synchronous) data format. The UART's output is applied to a *modem*, which is an acronym for modulator/demodulator. The serial pulses entering the modem are outputed as different frequency tones that are coupled to the telephone line either directly or by acoustic coupling with the telephone handset. At the receive end, the process is reversed. These systems are usually *half-duplex operation*, in that transmission can occur in either direction but not in both directions simultaneously. This is in contrast with a *full-duplex* system, such as a normal telephone conversation, where simultaneous transmission in both directions is possible. While the system shown in Fig. 8-22 connects two computers, many other schemes are possible. It may connect a computer to a computer terminal, two terminals, or a digital system of remote environmental sensors to a central monitoring computer, as examples.

UART Operation

UARTs are available, in chip form, for about $10. A glance at the transmit section of a typical unit shown in Fig. 8-23 provides an indication of the complexities involved. In simplified fashion, the transmit function is achieved as follows:

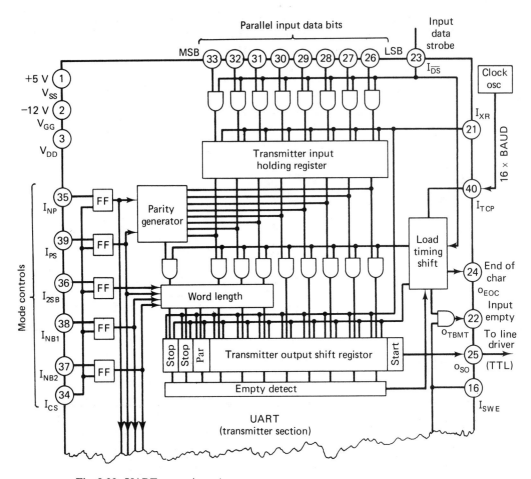

Fig. 8-23. UART transmit section.

1. The logic mode controls (pins 34 to 39) are applied either manually or by computer:
   A. Pin 36: This is the stop bit control. A logic 0 causes one stop bit to follow each serialized set of data bits in the UART output (pin 25). A logic 1 (+5V) causes two stop bits to follow the data.
   B. Pin 39: This is the parity select control. *Parity* pertains to the use of a self-checking code in which the total number of 1s (or 0s) in each code expression is always odd or even. A 0 at pin 39 checks for odd parity while a 1 checks even parity. If the number of bits per code expression is in error, a failure indication is provided at the receive UART.

    C. Pin 35: A 1 applied to this pin disables transmission of the parity bit and the receiver parity check and forces pin 13 to go to 0.

). Pins 37 and 38: These select the desired character length of 5, 6, 7, or 8 bits/character.

    Pin 34: A logic 1 causes the above-listed controls to be entered into the holding register.

r the above-mentioned controls are set up, the parallel data bits be entered into pins 26 to 33. The least significant bit (LSB) is d into pin 26 up to the most significant bit (MSB) at pin 33, if all its are used.

ata entry of step 2, a negative-going pulse is sent to pin 23 by the r after each character entry. It enters the parallel data into the lding register.

oscillator input at pin 40 allows the serial data output to ne data bit for every 16 clock pulses. In this way, the trans- e is controlled by the operator to suit the capabilities of the odem and telephone lines.

lata in the holding register is transferred to the transmitter gister, which actually performs the parallel-to-serial con-version.

The receiving section of the UART is similar to the transmitting section but functions in reverse order. The UART handles many complex operations on a single monolithic chip. It can be termed a microprocessor dedicated to a specific function, data handling. UARTs have added another impetus to the growth of data communications.

Modems

During the transmission, the UART's output is applied to the modem. The modem is functioning as a modulator. Low-speed modems use FSK by typically transmitting a 1070-Hz tone to represent a 0 and 1270 Hz for a 1. The tones are carried by the telephone line to the receiving modem. It acts as a demodulator by converting the FSK signal back to 1s and 0s. These serialized data are applied to the receive portion of a UART, which converts it to parallel data as required by the receiving computer. High-speed modems can transmit at data rates of up to 4800 bits/s by using multilevel phase modulation. One such system uses one of eight phase angles to represent three consecutive data bits. This process is illustrated in Fig. 8-24. Since each packet of three serial data bits provided to the transmit section of the modem have eight possible combinations, the eight different phase angles shown are capable of transmitting the necessary information. Since each phase shift implies a certain frequency shift, the receiving modem can be made capable of outputing the proper packet of three consecutive bits.

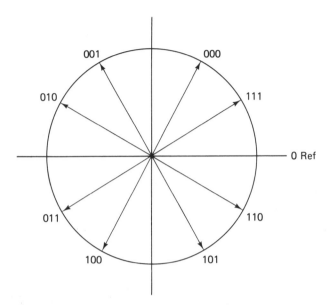

Fig. 8-24. Multilevel phase modulation.

Delay Equalizers

Since telephone lines are notorious for providing different amounts of delay to different frequencies, a *delay equalizer* is incorporated to minimize error rate. The delay-versus-frequency characteristic for a typical phone line is shown with dotted lines in Fig. 8-25, while the characteristic after equalization is also provided. The delay equalizer is a complex *LC* filter that provides increased delay to those frequencies least delayed by the phone line so that all frequencies arrive at nearly the same time.

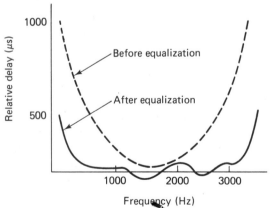

Fig. 8-25. Delay equalization.

8-6 RADIO TELEMETRY

Telemetry may be defined as remote metering. It is the process of gathering data on some particular phenomenon without the presence of human monitors. The gathered data may be recorded on chart recorders or tape recorders and then picked up at some convenient time. It may also be transmitted on a radio-wave carrier to a different location, in which case it is called *radio telemetry*. This process started during World War II, when telemetry systems were developed to obtain flight data from aircraft and missiles. It offered an alternative to having human observers on board when that was impractical or considered to be too dangerous.

Since situations to be remotely metered invariably involve more than one measurement, the different signals are always multiplexed. This allows the use of a single transmitter/receiver and, in fact, was the first major use of multiplexing techniques. It was also the beginning of pulse modulation techniques because of the ease with which they can be multiplexed.

Telemetry Block Diagram

A radio telemetry system block diagram is shown in Fig. 8-26. The process begins with the system to be monitored. Transducers (sensors) convert from the entity to be measured to an electrical signal. Five transducer outputs are shown in Fig. 8-26, but sophisticated systems, such as for a Mars exploration probe, may include hundreds of transmitted measurements. In any measurement system,

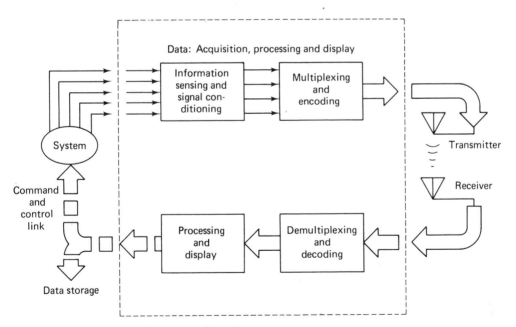

Fig. 8-26. Radio telemetry block diagram.

263

stimuli such as temperature, pressure, movement, or acceleration must be converted to a form that can be processed electrically. While a mercury thermometer produces a good visual output of temperature, a thermistor or thermocouple transducer converts temperature to an electrical signal usable by a telemetry system.

Following the sensing, Fig. 8-26 shows a signal conditioning function. The outputs from the transducers will be electrical in nature but may not have much else in common. The outputs may be variable resistance, capacitance, inductance, voltage, or current of all different magnitudes. The conditioning circuits turn the raw data into uniform digestable information. Following conditioning, the signals are applied to the multiplexing and encoding block. They are then transmitted to

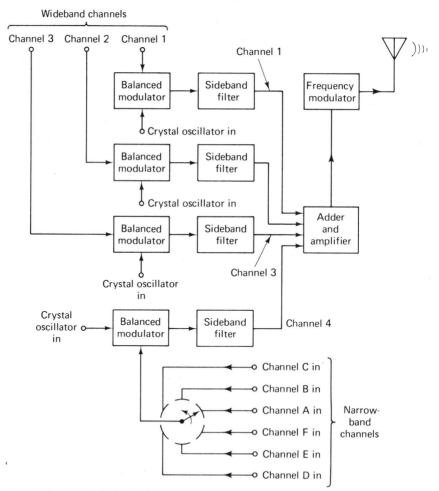

Fig. 8-27. TDM and FDM telemetry transmitter system.

the receiver for demultiplexing and decoding. It is followed by the processing and display function. Complex telemetry systems often incorporate a computer for maximum efficiency in data processing. From this block a command and control link is shown in Fig. 8-25 for those systems that send control signals back to the system under measurement. This may be done to change the performance to a desired level. This "completes the loop" except for the path to data storage shown in Fig. 8-26.

Typical Telemetry System

Telemetry systems may use FDM or TDM and in some cases both, as shown in Fig. 8-27. In this instance, three wide-band channels and six narrow-band channels are provided. Frequently, there are some conditions that must be monitored more often than others. They are represented by channels 1, 2, and 3 in Fig. 8-27. The six conditions that do not require as-rapid monitoring (subchannels A through F) are time-division-multiplexed onto channel 4. They are referred to as subchannels, since six of them make up one of the frequency-division-multiplexed main channels. There are countless modulation combinations possible in radio telemetry. In the system illustrated in Fig. 8-27, the subchannels are PCM-encoded to amplitude-modulate a subcarrier, which is converted into SSB and used to frequency-modulate the main carrier. The three wide-band channels use SSB directly to frequency-modulate the main carrier. This is termed a PCM/SSB/FM system. However, any other method is possible (and has probably been used), such as PAM/FM/FM, PDM/AM/FM, and PPM/AM/AM.

9

TELEVISION

9-1 INTRODUCTION

Television is a field of electronic technology that has more direct effect on the people of the world than any other. It is a very specialized branch of our technology that utilizes many of the principles already explained, as well as many new ones.

As a practitioner of electronics you have probably already been approached by friends or relatives with a request to fix a TV set. Not wanting to show any ignorance in your field of study, you leap into the breach only to discover that you do not have the skills (and perhaps specialized test equipment) necessary to make the repair. It should be comforting to you, however, that because of the extensive technology involved in TV, many experienced electronic engineers refuse to even attempt the repair of their TVs.

The concept of television was developed in the 1920s, feasibility was shown in the 1930s, commercial broadcasting started in the 1940s, and the ensuing years have seen the mushrooming growth of an industry so far-reaching that some sociologists make the study of its effects their life's work. The technology, while still undergoing continued improvements, has reached a certain level of maturity.

Today's 25-inch color TV that can be purchased for $600 contains the complexity of industrial equipment selling for $5000. Mass production techniques utilized in TV assembly enable the average consumer to afford this truly sophisticated piece of electronics.

9-2 TRANSMITTER PRINCIPLES

A TV transmitter is actually two separate transmitters. The *aural* or sound transmitter is actually an FM system very similar to a broadcast FM station. The system is still a "high-fidelity" system, since the same 30-Hz to 15-kHz audio range is transmitted. The major difference between broadcast FM and TV audio systems is that TV uses a ± 25-kHz deviation to conserve bandwidth, while you will recall broadcast FM uses a ± 75-kHz deviation. Thus, the TV aural signal has the same fidelity but is less effective in canceling the indirect noise effects explained in Chapter 5. It is ironic that most TV receivers utilize such poor sound systems, since musical broadcasts could be greatly enhanced with a high-fidelity audio amplifier and speaker system.

The *video* or "picture" signal is amplitude-modulated onto a carrier. Thus, the composite transmitted signal is a combination of both AM and FM principles. This is done to minimize interference effects between the two at the receiver, since an FM receiver is relatively insensitive to amplitude modulation and an AM receiver has rejection capabilities to frequency modulation.

Figure 9-1 shows a simplified block diagram for a TV system. The TV camera converts a visual "picture" or scene into an electrical signal. The camera is thus a transducer between light energy and electrical energy. At the receiver, the CRT picture tube is the analogous transducer that converts the electrical energy back into light energy.

The microphone and speaker shown in Fig. 9-1 are the similarly related

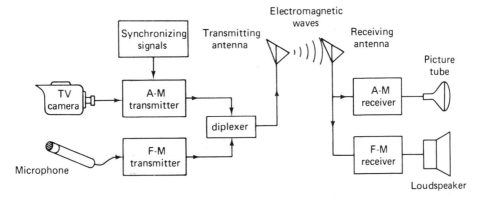

Fig. 9-1. Simplified TV system.

transducers for the sound transmission. There are actually two more transducers shown, the sending and receiving antennas. They convert between electrical energy and the electromagnetic energy required for transmission through the atmosphere.

The *diplexer* shown in Fig. 9-1 feeding the transmitter antenna feeds both the visual and aural signals to the antenna while not allowing either to be fed back into the other transmitter. Without the diplexer, the low output impedance of either transmitter power amplifier would dissipate much of the output power of the other transmitter. The synchronizing signal block will be explained in Sec. (9-3).

TV Cameras

The TV camera is optically focused so that the scene to be transmitted appears on its light-sensitive area. The early TV cameras were the *iconoscope* and the *image orthicon*. The most widely used type today is the *vidicon*. It has comparable performance to its forerunners but is much less costly and more compact in size. Figure 9-2 shows a sketch of a typical vidicon. Notice the three very thin layers at the tube's front surface. These are the light-active areas. The first is a transparent conductive film. Then a semiconductor photoresistive layer is deposited on the conductive film. A photoconductive *mosaic* layer is then deposited on the semiconductor layer. The middle semiconductor layer exhibits very high resistance when dark, but it reduces greatly when struck by photons of light. The photoconductive mosaic has around 1 million individual separate areas which act as tiny capacitors, with the photoresistive layer acting as the dielectric and the conductive film as the other common electrode. The dielectric leakage is thus variable and dependent on the amount of light striking each area (capacitor).

The electron beam scans across (and slowly down) the mosaic areas so as to charge up each of the many tiny capacitors. Light on the mosaic areas discharges

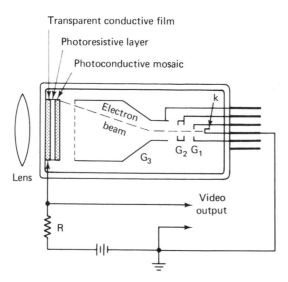

Fig. 9-2. Vidicon camera tube.

the capacitors through the load resistor *R* in Fig. 9-2. The scanning electron beam (developed by the cathode *K* and three grids) recharges the mosaic capacitors and produces a video signal voltage drop across *R* that is proportional to the light intensity at the individual areas being scanned. Thus, the video output is a signal whose instantaneous output is the result of scanning just one of the tiny capacitors at a time.

Scanning

To understand how these tiny individual outputs can serve to represent an entire scene, refer to Fig. 9-3. In this simplified system the camera focuses the letter "T" onto the capacitors of the vidicon, but instead of 1 million capacitors, this system has just 30, arranged in six rows with 5 capacitors per row.

The letter "T" is focused on to the light-sensitive area such that all of rows 1 and 6 are illuminated [Fig. 9-3(a)], while all of row 2 is dark and the centers of rows 3, 4, and 5 are dark. Now, if the electron beam is made to scan each row sequentially and if the *retrace* time is essentially zero, then Fig. 9-3(b) shows the sequential

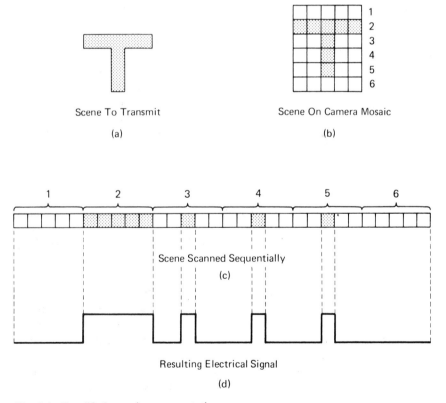

Fig. 9-3. Simplified scanning representation.

breakup of information. The retrace interval is the time it takes the moving electron beam to move from the end of one line back to the start of the next lower line. It is usually accomplished very rapidly. The variable light on the photoresistive capacitor dielectric causes a variable capacitive discharge which results in a similar variable voltage being developed at the vidicon's output, as shown in Fig. 9-3(c). The visual scene has been converted to a video (electrical) signal and can now be suitably amplified and used to amplitude-modulate a carrier for broadcast.

The picture for broadcast TV has been standardized at a 4:3 ratio of the width to height. This is termed the *aspect ratio* and was selected as the most pleasing picture size to the human eye.

9-3 *TRANSMITTER/RECEIVER SYNCHRONIZATION*

When the video signal is detected at the receiver, some means of *synchronizing* the transmitter and receiver is necessary:

1. When the TV camera starts scanning line 1, the receiver must also start scanning line 1 on the CRT output display. You do not want the top of a scene appearing at the center of the TV screen.
2. The speed that the transmitter scans each line must be duplicated by the receiver scanning process to avoid distortion in the receiver output.
3. The *horizontal retrace* or time when the electron beam is returned back to the left-hand side to start tracing a new line must occur coincidentally at both transmitter and receiver. You do not want your horizontal lines starting at the center of your TV screen.
4. When a complete set of horizontal lines has been scanned, moving the electron beam from the end of the bottom line to the start of the top line (vertical flyback or retrace) must occur simultaneously at both transmitter and receiver.

It is thus seen that visual transmissions are more complex than audio because of these synchronization requirements. At this point, voice transmission seems elementary because it can be sent on a continuous basis without synchronization.

Thus, the other major function of the transmitter besides developing the video and audio signals is to generate synchronizing signals that can be used by the receiver so that it stays in step with the transmitter.

In the scanning process, the electron beam for both transmitter and receiver starts at the upper left-hand corner and sweeps horizontally to the right side. It then is rapidly returned to the left side, and this interval is termed *horizontal retrace*. An appropriate analogy to this process is the movement of your eye as you read this line and rapidly retrace to the left and slightly drop for the next line. When all the horizontal lines have been traced, the electron beam must move from the lower right-hand corner up to the upper left-hand corner. This *vertical retrace interval* is analogous to the time it takes the eye to move from the bottom of one page to the top of the next.

FCC regulations stipulate that U.S. TV broadcasts shall consist of 525 horizontal scanning lines. Of these, about 40 lines are lost as a result of the vertical retrace interval. This leaves 485 visible lines which you can actually see if a TV screen is viewed at close range. The number of visible lines does not depend on the TV screen size.

Because this scanning occurs rapidly, persistence of vision and CRT phosphor persistence causes us to perceive these 485 lines as a complete image.

Interlaced Scanning

The *frame frequency* is the number of times per second that a complete set of 485 lines (complete picture) is traced. That rate for broadcast TV is 30 times per second. Stated another way, a complete scene (frame) is traced every $\frac{1}{30}$ s. Thirty frames per second is not enough to keep the human eye from perceiving *flicker* as a result of a noncontinuous visual presentation. This flicker effect is observed when watching old-time movies. If the frame frequency were increased to 60 per second, the flicker would no longer be apparent, but the video signal bandwidth would have to be double. Instead of that solution, the process of *interlaced scanning* is used to "trick" the human eye into thinking it is seeing 60 pictures per second.

Figure 9-4 illustrates the process of interlaced scanning. The first set of lines (the first *field*) is traced in $\frac{1}{60}$ s, and then the second set of lines (the second *field*) that comprises a full scene (485 lines total) is interleaved between the first lines in the next $\frac{1}{60}$ s. Therefore, lines 2, 4, 6, etc., occur during the first field, with lines 1, 3, 5, etc., interleaved between the even-numbered lines. The field frequency is thus 60 Hz, with a frame frequency of 30 Hz. This illusion is enough to convince the eye that 60 pictures per second occur when, in fact, there are only 30 full pictures per second.

The process of interlacing in TV is analogous to a trick used in motion picture projection to prevent flicker (noncontinuous motion). In motion pictures the goal is to conserve film rather than bandwidth, and this is accomplished by flashing each of the 24 frames per second onto the screen twice so as to create the illusion of 48 pictures per second.

Horizontal Synchronization

To accommodate the 525 lines (485 visible) every $\frac{1}{30}$ s, the transmitter must send a synchronization (sync) pulse between every line of video signal so that perfect transmitter–receiver synchronization is maintained. The detail of these pulses is shown in Fig. 9-5. Three horizontal sync pulses are shown along with the video signal for two lines. The actual horizontal sync pulse rides on top of a *blanking pulse*, as shown in the figure. The blanking pulse is a strong enough signal so that the electron beam retrace at the receiver will be "blacked" out and thus invisible to the viewer. The interval before the horizontal sync pulse appears on the blanking

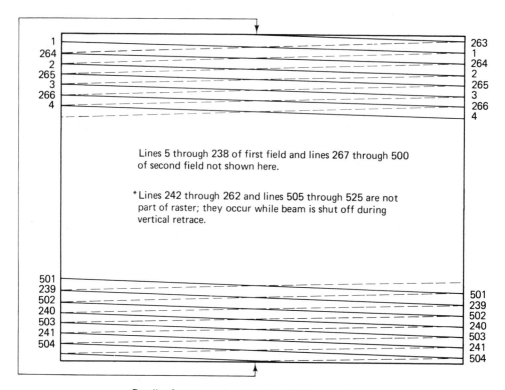

Details of raster produced by the 525-line scanning pattern

Fig. 9-4. Interlaced scanning.

pulse is termed the *front porch*, while the interval after the end of the sync pulse, but before the end of the blanking pulse, is called the *back porch*. Notice in Fig. 9-5 that the back porch includes an eight-cycle sine-wave burst at 3,579,545 Hz. It is appropriately called the *color burst*, as it is used to calibrate the receiver color subcarrier generator. Further explanation on it will be provided in Sec. (9-12). Naturally enough, a black-and-white broadcast does not include the color burst.

The two lines of video picture signal shown in Fig. 9-5 can be described as follows:

> *Line 2:* It starts out nearly fully black at the left-hand side and gradually lightens to full white at the right-hand side.
> *Line 4:* It starts out medium gray and stays there until one-third of the way over it gradually becomes black at the picture center. It suddenly shifts to white and gradually turns darker gray at the right-hand side.

Since the horizontal sync pulses occur once for each of the 525 lines every $\frac{1}{30}$ s, the frequency of these pulses will be

$$525 \times 30 = 15.75 \text{ kHz}$$

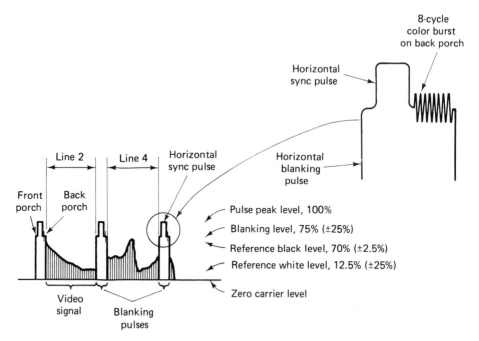

Fig. 9-5. Horizontal sync pulses.

Thus, both transmitter and receiver must contain 15.75-kHz *horizontal* oscillators to control horizontal electron beam movement.

Vertical Synchronization

The vertical retrace and thus vertical sync pulses must occur after each $\frac{1}{60}$ s since the two interlaced fields that make up one frame (picture) occur 60 times per second. The video signal just before, during, and after vertical retrace is shown in Fig. 9-6. Notice that two horizontal sync pulses and the last two lines of video information of a field are initially shown. These are followed in succession by:

1. Equalizing pulses at a frequency double the horizontal sweep rate or 15.75 kHz × 2 = 31.5 kHz. They each have a duration of about 2.7 μs, with a period of 1/31.5 kHz or 31.75 μs. They are used to keep the receiver horizontal oscillator in sync during the relatively long (830 to 1330 μs) vertical blanking period.
2. One vertical sync pulse with a 190-μs pulse width and five serrations having a duration of 4.4 μs at 27.3-μs intervals. These serrations are used to keep the horizontal oscillator synchronized during the vertical sync pulse interval.
3. More equalizing pulses.
4. Horizontal sync pulses until the entire vertical blanking period has elapsed.

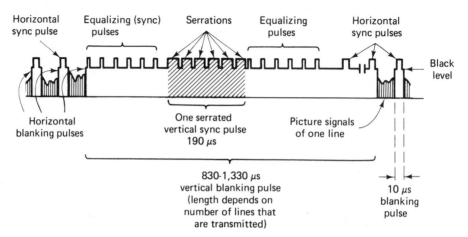

Fig. 9-6. Vertical retrace interval video signal.

Notice that the vertical blanking period is variable since the number of visible lines transmitted can vary between 482 and 495 at the discretion of the station. All other aspects of the pulses such as number, width, and rise and fall times, are tightly specified by the FCC so that all receiver manufacturers will know precisely what type of signals their sets will have to process. Notice also that the color burst necessary for all color transmissions is not shown in Fig. 9-6. When it is present, recall that it occurs on the back porch of each horizontal sync pulse.

The vertical sync pulses occur at a frequency of 60 Hz, which is the same frequency as the ac line voltage in the United States. This allows for good stability of the vertical oscillator and minimizes interference in the receiver. In Europe, where 50-Hz line voltage exists, a 50-Hz vertical oscillator system is used.

9-4 RESOLUTION

To provide adequate resolution, the video signal must include modulating frequency components from dc up to 4 MHz. This requires a truly wide-band amplifier, and amplifiers that have bandpass characteristics from dc up into the MHz region have come to be known as *video amplifiers*.

Resolution is the ability to resolve detailed picture elements. We already have an idea as to resolution in the vertical direction. Since about 485 separate horizontal lines are traced per picture, it might seem that the vertical resolution would be 485 lines. *Vertical resolution* may be defined as the number of horizontal lines that can be resolved. However, the actual resolution turns out to be about 0.7 of the number of horizontal lines, or

$$0.7 \times 485 = 339$$

Thus, the vertical resolution of broadcast TV is about 339 lines.

Horizontal resolution is defined as the number of vertical lines that can be resolved. A little mathematical analysis will show this capability. The maximum modulating frequency has already been stated as 4 MHz. The more vertical lines to resolve, the higher will be the frequency of the resulting video signal. The horizontal trace occurs at a 15.75-kHz frequency, and thus each line is 63.5 μs (1/15.75 kHz) in duration. The horizontal blanking time is about 10 μs leaving 53.5 μs. Since two lines (a white and a black) can be converted into a cycle of video signal, the number of vertical lines resolvable is

$$4 \text{ MHz} \times 53.5 \ \mu s \times 2 = 428$$

Thus, the horizontal resolution is about 428 lines. Note that the 428 vertical lines conform nicely to the 339 horizontal lines when one remembers that a TV screen has a 4:3 width to height (aspect) ratio (428/339 $\simeq$ 4/3). Thus, equal resolution exists in both directions, as is desirable. The reader should recognize that increased modulating signal rates above 4 MHz would allow for increased vertical or horizontal resolutions or some increase for both. This is shown in the following examples.

Example 9-1

Calculate the increase in horizontal resolution possible if the video modulating signal bandwidth were allowed up to a 5 MHz maximum.

Solution:

The 53.5 μs allocated for each visible trace could now develop a maximum 5-MHz video signal. Thus, the total number of vertical lines resolvable is

$$53.5 \ \mu s \times 5 \text{ MHz} \times 2 = 535 \text{ lines}$$

Example 9-2

Determine the possible increase in vertical resolution if the video frequency were allowed up to 5 MHz.

Solution:

The visible horizontal trace time can now be decreased if the horizontal resolution can stay at 428 lines. That new trace time is

$$\text{trace time} \times 5 \text{ MHz} \times 2 = 428$$

$$\text{trace time} = 42.8 \ \mu s$$

Once again, assuming that 10 μs is used for horizontal blanking, that means 52.8 μs total can be allocated for each horizontal trace. With $\frac{1}{30}$ s available for a full picture, that implies a total number of horizontal traces of

$$\frac{\frac{1}{30} \text{ s}}{52.8 \ \mu\text{s}} = 632 \text{ lines}$$

Allowing 32 lines for vertical retrace means a vertical resolution of

$$600 \times 0.7 = 420 \text{ lines}$$

The previous two examples are excellent proofs of Hartley's law. It is very plain to see that an increase in bandwidth led to the possibility of greater transmitted information (in the form of increased resolution).

9-5 THE TELEVISION SIGNAL

The maximum modulating signal for the video signal is 4 MHz. Since it is amplitude-modulated onto a carrier, a bandwidth of 8 MHz is implied. However, the FCC allows only a 6-MHz (*only* is a relative term here, since 6 MHz is enough to contain 600 AM radio broadcast stations of 10 kHz each) bandwidth per TV station and that must also include the FM audio signal. The TV signal that is transmitted is shown in Fig. 9-7.

The lower visual side band extends only 1.25 MHz below its carrier with the remainder filtered out, but the upper side band is transmitted in full. The audio

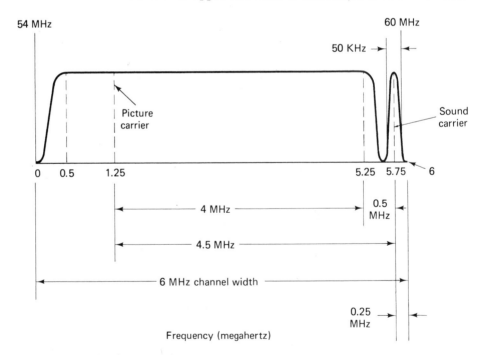

Fig. 9-7. Transmitted TV signal (Channel 2).

carrier is 4.5 MHz above the picture carrier, with its FM side bands fully transmitted up to its ± 25 kHz deviation limit. The 54- to 60-MHz limits shown in Fig. 9-7 are the allocation for channel 2. The table in Fig. 9-8 shows the complete allocation for all the VHF and UHF channels. Notice that the VHF channels are broken up into two bands—54 to 88 MHz and 174 to 216 MHz. The UHF band is continuous and eats up a tremendous chunk of the usable frequency spectrum, as can be seen.

Lower VHF Band		Upper VHF Band		UHF Band	
Channel	Lowest frequency, MHz	Channel	Lowest frequency, MHz	Channel	Lowest frequency, MHz
2	54	7	174	14	470
3	60	8	180	24	530
4	66	9	186	34	590
	(4 MHz skipped)	10	192	44	650
5	76	11	198	54	710
6	82	12	204	64	770
		13	210	74	830
				83 (highest)	884

Fig. 9-8. TV channel allocations.

The lower side band is mostly removed by filters that occur near the transmitter output. While only one side band is necessary, it would be impossible to filter out the entire lower side band without affecting the amplitude and phase of the lower frequencies of the upper side band and the carrier. Thus, part of the 6-MHz bandwidth is occupied by a "vestige" of the lower side band. It is therefore commonly referred to as *vestigial-side-band* operation. It offers the added advantage that carrier reinsertion at the receiver is not necessary as in SSB, since the carrier is not attenuated in vestigial-side-band systems.

Once the entire TV signal is generated, it is amplified and driven into an antenna that converts the electrical energy into radio (electromagnetic) waves which travel through the atmosphere to be intercepted by a TV receiving antenna and fed into the receiver once again as an electrical signal. That signal consists of the video, audio, and synchronizing signals. The synchronizing signals are contained in the video signal, as previously shown.

9-6 TELEVISION RECEIVERS

A TV receiver utilizes the superheterodyne principle, as do virtually all other types of receivers. It does become a bit more complex than most others because it must handle video and synchronizing signals as well as the audio that previously studied receivers do. A block diagram for a typical TV receiver is shown in Fig. 9-9.

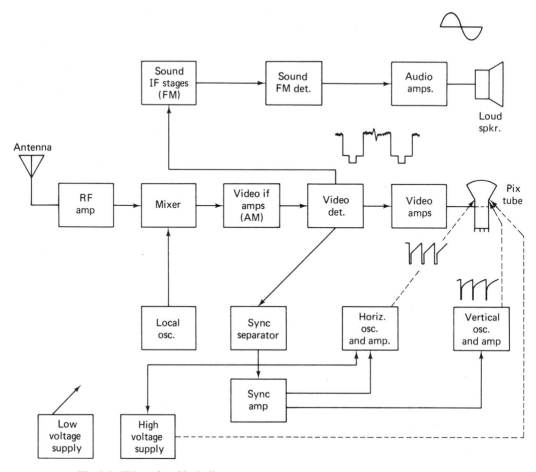

Fig. 9-9. TV receiver block diagram.

The incoming signal is selected and amplified by the RF amp and stepped down to the IF frequency by the mixer–local oscillator blocks. The IF amplifiers handle the composite TV signal, and then the video detector separates the sound and video signals. The sound signal detected out of the video detector is the FM signal that is sent into the sound channel block, which is a complete FM receiver system in itself. The other video detector output is the video (plus sync) signal. The actual video portion of the video signal is amplified in the video amp and subsequently used to control the strength of the electron beam that is scanning the phosphor of the CRT. The sync separator separates the horizontal and vertical sync signals, which are subsequently used to precisely and periodically calibrate the horizontal and vertical oscillators. The oscilllator outputs are then amplified and used to precisely control the horizontal and vertical movement of the electron beam that is scanning the phosphor of the CRT. They are applied to a coil around the yoke of the tube whose magnetic fields cause the electron beam to be deflected in the proper fashion. This coil around the tube yoke is commonly referred to as the yoke.

The low-voltage power supply shown in the receiver block diagram is used to power all the electronic circuitry. The high-voltage output is derived by stepping up the horizontal output signal (15.75 kHz) via transformer action. This transformer, usually termed the *flyback transformer*, has an output of 10 kV or more, which is required by the CRT anode to make the electron beam travel from its cathode to the phosphor.

In the following sections we shall provide greater detail on the basic operation just presented.

9-7 THE FRONT END

The front end of a TV receiver is also called the *tuner* and contains the RF amplifier, mixer, and local oscillator. Its output is fed into the first IF amplifier. A pictorial block diagram for the tuner is shown in Fig. 9-10. Since the FCC requires all sets to contain provisions for UHF as well as VHF tuning, Fig. 9-10 shows two different inputs. It is the obvious function of the tuner to select the desired station and to reject all others, but these important functions are also performed:

1. It selects the desired station.
2. It provides amplification.

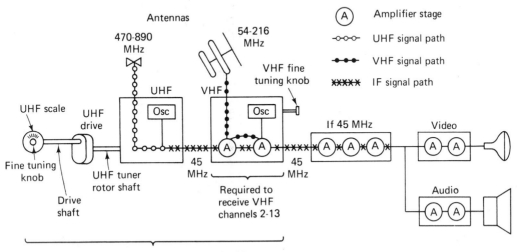

Fig. 9-10. A UHF-VHF tuner block diagram. (Courtesy of Zenith Radio Corp.)

3. It prevents the local oscillator signal from being driven into the antenna and thus radiating unwanted interference.

4. It steps the received RF signal down to the frequency required for the IF stages.

5. It provides proper impedance matching between the antenna-feed line combination into the tuner itself. This allows for the largest possible signal into the tuner and thus the largest possible signal-to-noise ratio.

VHF Tuners

There are three basic VHF tuner types. The *turret* or *drum* type is shown in Fig. 9-11. It is so named because the coils used to tune in different stations are contained in a drum-type structure that is rotated by the tuner selector knob. When a station is selected, the coils for the required RF amp, mixer, and local oscillator frequencies are swung into position so that contact is made into their respective circuits via fixed spring terminals.

Fig. 9-11. Turret or "Drum" tuner.

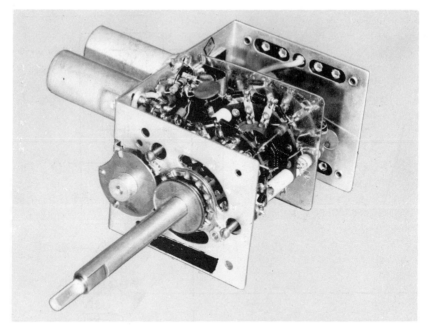

Fig. 9-12. A wafer-switch tuner. (From Matthew Mandl, *Modern Television Systems: Theory and Servicing,* © 1974, Prentice-Hall, Inc., p. 96.)

A second type of tuner is the *wafer-switch* variety, shown in Fig. 9-12. A series of wafer switches allows selection of the proper RF, mixer, and local oscillator coils. The coils are wound around the outer rim of the switch. The coils contain several turns for channels 2 through 6, with single or partial turns for channels 7 through 13 providing the necessary resonant frequencies. A typical VHF tuner schematic using a wafer switch is shown in Fig. 9-13. The turret tuner could also be used with this circuitry. The RF amplifier stage uses the common-base configuration because of its better voltage-gain performance at high frequencies. Notice that its gain is controlled by an AGC input. The RF output is coupled into the mixer stage via magnetic coupling between the RF and mixer coils. The mixer and local oscillator stages are both common-emitter configurations. Notice the variable inductor in series with the individual oscillator coils. This is the fine-tuning control available to the set operator for precise tuning of each individual station. The AFC input is a variable dc bias that provides an automatic frequency control to compensate for minor drifts in the oscillator frequency. The diode shown is reverse-biased and exhibits a variable capacitance based on the amount of reverse bias applied at the AFC feedback input. That capacitance change provides slight adjustment to the frequency of oscillation.

The third major type of VHF tuner is the *varactor* or *pushbutton* or *electronic* tuner. It is being used increasingly on newer receivers. The varactor diode is spe-

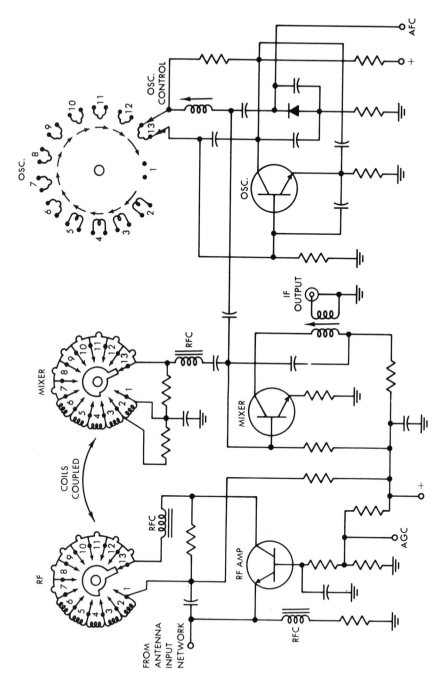

Fig. 9-13. VHF tuner schematic. (From Matthew Mandl, *Modern Television Systems: Theory and Servicing,* © 1974, Prentice-Hall, Inc., p. 96.)

282

cially constructed to emphasize the capacitance-versus-reverse bias characteristic of a standard diode. Recent advances in varactor diode fabrication techniques have made this type of tuner practical. Figure 9-14 shows a schematic for a varactor tuner. A voltage divider is set up, and the reverse bias for each of the four varactor diode tuned circuits is determined by which pushbutton is closed. The figure shows channel 4 tuned in. The need for a well-regulated dc power supply is obvious, since variations would cause detuning of the desired station.

The VHF tuner is required to tune from 54 to 216 MHz. This is too wide a band for the amount of capacitance variation possible from the varactor diodes.

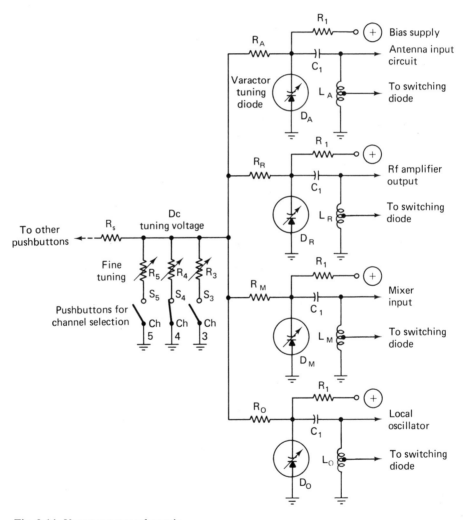

Fig. 9-14. Varactor tuner schematic.

Consequently, when the high band (channels 7 through 13) is being selected, a switching diode shorts the bottom portion of the tuned circuit's inductance to ground, providing a band switch capability.

A fourth type of tuner is the PLL synthesizer. As PLL technology is improved and costs go down, more tuners of this type will be incorporated.

UHF Tuners

A reference to Fig. 9-10, which showed a pictorial block diagram of a combination UHF-VHF tuner, is helpful at this point. Notice the output of the UHF block, which is already at the indicated IF frequency of 45 MHz. That signal is usually developed by mixing the incoming UHF signal with an oscillator signal through the nonlinearity of a diode. The process is illustrated in Fig. 9-15. There is no RF amp or transistor-type mixer, as you would probably expect. Instead, the UHF signal is immediately stepped down to the IF frequency and then goes to the VHF RF amp and mixer, which provide the gain which makes the UHF and VHF signals of about equal strength when they go into the IF amps. This is done because RF amps and transistor mixers (that supply gain) tend to be more costly and also offer relatively poor signal-to-noise ratios at the high UHF frequencies.

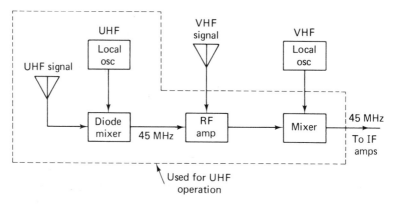

Fig. 9-15. UHF tuner block diagram.

The local oscillator for UHF tuners can be a mechanically variable inductor or capacitor or, in some more recent sets, a varactor diode arrangement that allows a number of specific UHF stations to be selected with pushbuttons.

9-8 IF AMPLIFIERS

The IF amplifier section is fed from the mixer output of the tuner. It is often referred to as the video IF, even though it is also processing the sound signal. Sets that process the sound and video in the same IF stages are known as *intercarrier* systems. Very early sets used completely separate IF amps for the sound and video signals. The IF stages of intercarrier sets are often referred to as the video IF, even though they also handle the sound signal because the sound signal is also processed by another IF stage after it has been extracted from the video signal. From now on the video IF will be referred to simply as the IF.

The major functions of the TV IF stage are the same as in a regular radio receiver: to provide the bulk of the set's selectivity and amplification. The standard IF frequencies are 45.75 MHz for the picture carrier and 41.25 MHz (45.75 MHz − 4.5 MHz) for the sound carrier. Recall that mixer action causes a reversal in frequency when the IF amplifier accepts the difference between the higher local oscillator frequency and the incoming RF signal. Therefore, the sound carrier that is 4.5 MHz above the picture carrier in the RF signal ends up being 4.5 MHz below it in the mixer output into the first IF stage. The inversion effect of IF frequencies when receiving channel 5 are shown in Fig. 9-16. The IF frequencies are always equal to the difference between the local oscillator and RF frequencies.

Channel 5 76-82 MHz	Transmitted RF Frequency, MHz	Local Oscillator Frequency, MHz	IF Frequency MHz
Upper channel frequency	82 MHz	123 MHz	41 MHz
Sound carrier	81.25	123	41.25
Picture carrier	77.25	123	45.75
Lower channel frequency	76	123	47

Fig. 9-16. IF signal frequency inversion.

Stagger Tuning

A major difference between radio and TV IF amps is that most radio receivers require relatively high-Q tuned circuits, since the desired bandwidth is often less than 10 kHz. A TV IF amp requires a pass band of about 6 MHz because of the wide frequency range necessary for video signals. Hence, the problem here is not how to get a very narrow bandwidth with high-Q components but how to get a wide-enough bandwidth but still have relatively sharp falloff at the pass-band edges. Most TV IF amps solve this problem through the use of *stagger* tuning. Stagger tuning is the technique of cascading a number of tuned circuits with slightly different resonant frequencies, as shown in Fig. 9-17. The response of three separate *LC* tuned circuits is used to obtain the total resultant pass band shown with dotted lines. The use of a lower-Q tuned circuit in the middle helps provide a flatter overall response than would otherwise be possible.

IF Amplifier Response

The ideal overall IF response curve in Fig. 9-18 provides some interesting food for thought. The sound IF carrier and its narrow side bands are amplified at only one-tenth the midband IF gain. This is done to minimize interference effects that the sound would otherwise have on the picture. You may have noticed a TV with normal picture when no audio is present but with visual interference in step with the sound output. This is an indication that the sound signal in the IF is not

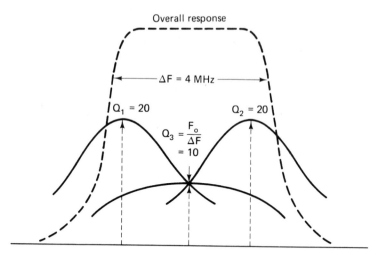

Fig. 9-17. Illustration of stagger tuning.

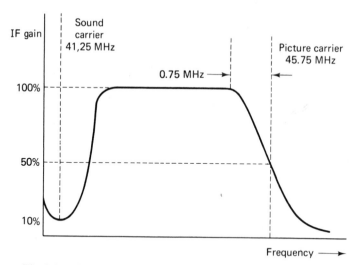

Fig. 9-18. Ideal IF response curve.

attenuated enough and can often be remedied by adjustment of the set's fine-tuning control.

Another interesting point illustrated in Fig. 9-17 is the attenuation given to the video side frequencies right around the picture carrier. This is done to reverse the vestigial-side-band characteristic generated at the transmitter. Refer back to Fig. 9-7 to refresh your memory on the transmitted characteristic. If the receiver

IF response were equal for all the video frequencies, the lower ones (up to 0.75 MHz) would have excessive output because they have both upper-and lower-sideband components.

Wavetraps

To obtain the steep attenuation curve for the sound carrier shown in Fig. 9-18 it is necessary to incorporate a *wavetrap* or, more simply, *trap*, in the IF stage. A trap is a high-*Q bandstop* circuit that attenuates a narrow band of frequencies. It can be a series resonant circuit that shorts a specific frequency to ground, as in Fig. 9-19(a), or a parallel resonant circuit that blocks a specific frequency, as in

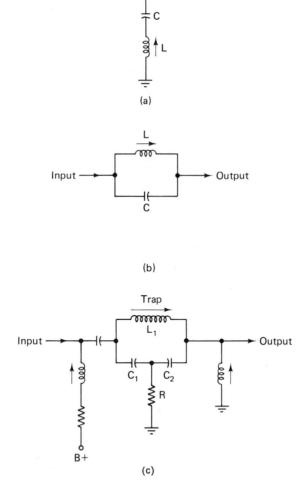

Fig. 9-19. Wave traps.

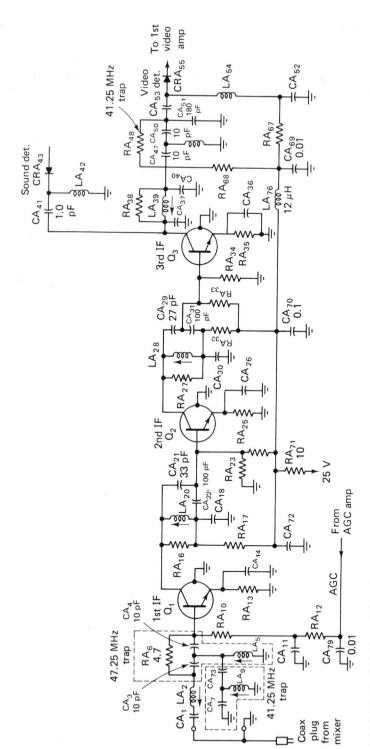

Fig. 9-20. Typical IF amplifier schematic.

288

Fig. 9-19(b). Even greater attenuation of a specific frequency is obtained with the *bridged-T* trap in Fig. 9-19(c). Traps are also employed in quality sets to eliminate carrier signals of adjacent channels. The carrier signal of an upper adjacent channel occurs at 39.75 MHz. While adjacent channels are not assigned in the same city, it is possible for a location midway between two adjacent channel stations to receive severe interference without a 39.75-MHz trap. A similar problem can exist with the sound carrier of a lower adjacent channel, which would occur at 47.25 MHz.

Typical IF Amplifier

A typical IF amp schematic is provided in Fig. 9-20. At the input are traps for 41.25 and 47.25 MHz. Following the traps, the IF signal is fed into the first transistor, Q_1, a common-emitter stage with a bypassed emitter resistor to provide high voltage gain. An AGC voltage applied into Q_1's base controls its gain. The following two IF stages (Q_2 and Q_3) are similar except they are not controlled by an AGC level. The tank circuits in all three collector (output) circuits are resonant at slightly different frequencies, providing the desired stagger tuning. An additional trap to the sound carrier (41.25 MHz) is provided just prior to the video detect diode. The output of the video detect diode is the original video modulating signal that extends from 0 to 4 MHz. It is obtained by beating the video carrier component at 45.75 MHz and the video side bands extending from 0 to 4 MHz below the carrier through the nonlinear diode. The subsequent video amplifier accepts only the difference components that extend from 0 to 4 MHz.

Audio Detection

The output of the separate sound detector diode in Fig. 9-20 is the FM sound signal centered at a frequency of 4.5 MHz. It is developed by the mixing action between the sound signal centered at 41.25 MHz and the picture carrier at 45.75 MHz. Since

$$45.75 \text{ MHz} - 41.25 \text{ MHz} = 4.5 \text{ MHz}$$

a frequency selective network at 4.5 MHz following the sound detector diode then forms part of the separate sound IF. This process, utilized in *intercarrier* receivers, ensures that the sound IF will stay at precisely 4.5 MHz in spite of any drift in the TV's local oscillator. The block diagram in Fig. 9-21 shows the subsequent processing for the sound signal. Following the sound IF amplifier, any form of FM discriminator can be used, such as the Foster–Seely, ratio detector, phase-locked loop, or quadrature detection. The last two methods lend themselves to IC usage

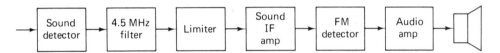

Fig. 9-21. Sound system block diagram.

because they eliminate the need for tuned circuits at a 4.5-MHz frequency. In some modern sets, the entire aural signal processing after an initial 4.5-MHz filter right up to the speaker is accomplished with a single special-purpose IC.

9-9 THE VIDEO SECTION

The function of the video section is outlined in the block diagram of Fig. 9-22. It takes the output of the video detector (0 to 4 MHz) and amplifies it to sufficient level to be applied to the picture tube cathode. Once applied to the cathode, this signal varies or *modulates* the electron beam strength such that white spots of a scene appear white and black spots appear black on the CRT face. It, of course, causes electron beam strengths of "in-between" magnitudes to provide various shades of gray.

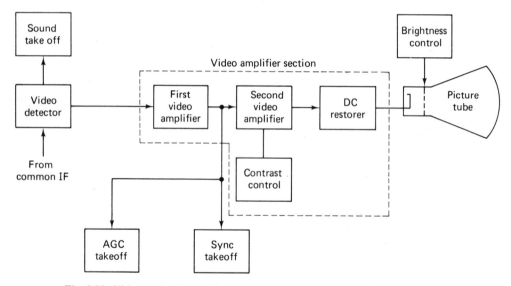

Fig. 9-22. Video section block diagram.

The contrast control block in Fig. 9-22 is analogous to the volume control of a radio receiver—it simply varies the amplitude of the signal applied to the CRT. The larger the difference in amplitude between maximum and minimum, the greater will be the picture contrast (difference between black and white).

The sync takeoff block in Fig. 9-22 is the point where the horizontal and vertical sync pulses are extracted from the video signal. In Sec. (9-10) we shall provide further elaboration on this subject. In some receivers the sync takeoff occurs after the final video amplifier stage. The sound takeoff may occur after several stages of video amplification rather than at the video detector, as shown in Fig. 9-22. In color sets, however, the sound takeoff must occur even before the video detector, as shown in the IF schematic of Fig. 9-20.

The *dc restoration* block in Fig. 9-22 is not necessary if the video amplifiers use direct coupling. If, however, capacitive coupling is used, as is usually more economical, then the dc portion of the video signal is lost. Without dc restoration the picture background levels will be erroneous, and in color sets their color will be incorrect.

The brightness control (Fig. 9-22) is a user adjustment, just as is the contrast control. The brightness control simply varies the dc level applied to the control grid or cathode of the CRT. It is *not* connected to the video signal amplitude in any way. It controls the overall picture brightness and *not* the min-max video signal level, as does the contrast control.

The video section also provides a takeoff point for the AGC signal. That signal is used to control the gain of previous amplifying stages such as the RF amp, mixer, and IF stages. This is necessary so that both strong and weak stations end up supplying the CRT cathode with approximately the same signal level and thus provide the same picture illumination. If the received signal is too weak, however, the electrical noise predominates over the desired signal and results in a "snowy" picture.

Typical Video Amplifier

Figure 9-23 shows a typical video amplifier section. Several points are worthy of note in this schematic.

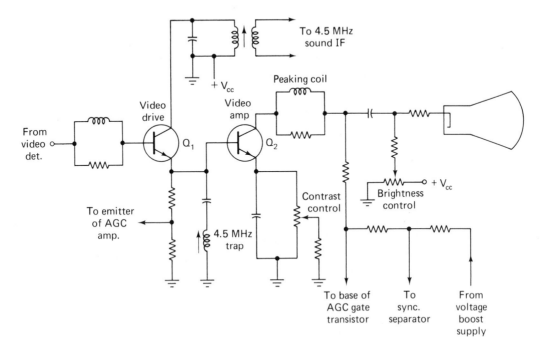

Fig. 9-23. Video section schematic.

1. The use of *peaking coils* is prevalent in video amplifiers. Since the video amps are required to function over such a wide frequency range, their high-frequency response falls off at the higher video frequencies without the peaking coils. The coil impedance becomes effective at the 3- to 4-MHz range and serves to increase the video amplifier's collector impedance, thus increasing its gain at the same time its interelectrode capacitance is decreasing gain. Thus, a relatively flat response is obtained up to 4 MHz.

2. Notice the 4.5-MHz trap at the output of the video drive transistor. Its function is to short out the rather strong intercarrier 4.5-MHz signal (recall that the picture carrier minus sound carrier equals 4.5 MHz). Of course, that signal, if it reached the CRT cathode, would cause severe picture interference. It is, however, the desired signal that is selected by the resonant circuit in the video driver's collector to be sent into the FM sound section.

3. This video section is direct-coupled and thus does not require dc restoration circuitry.

4. The gain of the video section is controlled by varying the amount of resistance in the video amplifier's emitter circuit. Recall that the gain of a common-emitter stage is inversely proportional to emitter resistance. This potentiometer is therefore the user contrast control.

5. The dc supply for the video amplifier transistor is obtained from the *voltage boost*. This is a special dc supply at a relatively high voltage level that enables a video drive signal for the CRT cathode of perhaps several hundred volts peak to peak. Thus, the dc voltage boost supply must be several hundred volts, and the video amplifier transistor must have very high voltage ratings.

9-10 SYNC AND VERTICAL DEFLECTION

The video section provides a takeoff point for the sync signals. Since the set needs both vertical (at 60 Hz) and horizontal (at 15.75 kHz) sync pulses, a means to separate one from the other is necessary. The *sync separator* is the circuit that performs this function. The key factor that enables separation is the fact that the vertical sync pulse is of long duration, while the horizontal sync pulse is of very short duration. In addition to separating one from the other, the sync separator *clips* the sync pulse off the video signal to prevent the sweep instability that could occur because of false synchronization of the sweep oscillators by spurious video signals. Because of this, the sync separator is sometimes referred to as the *clipper*.

Once the sync separator has clipped the sync signals from the lower-level video signal, the two types of sync pulses are applied to both low- and high-pass filters. The output of the low-pass filter will be the lower-frequency vertical sync pulse at 60 Hz since it is a wide pulse rich in low-frequency components. The output from the high-pass filter will be the horizontal sync pulse at 15.75 kHz since it is a very narrow pulse which is rich in high-frequency content. A low-pass filter is also termed an *integrator*, while a high-pass filter is classified as a *differentiator*.

This entire process of clipping and separation is shown in Fig. 9-24. The low-pass filter can simply be a shunt capacitance that shorts high frequencies to ground. Similarly, the high-pass filter includes a series capacitance that blocks low frequencies from reaching its output.

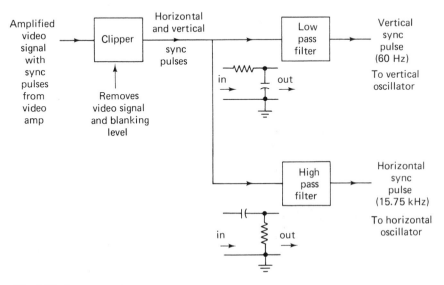

Fig. 9-24. Sync separator.

The vertical sync pulse is applied to the vertical oscillator. The vertical oscillator by itself generates a signal at *about* 60 Hz but must be at *precisely* the same frequency as the transmitter's vertical oscillator to prevent the picture from "rolling" in a vertical direction. The vertical adjustment control available to the set user adjusts the vertical oscillator's frequency to enable it to be brought into the range necessary so that it can "lock" onto the frequency of the sync pulse. The foregoing discussion also applies to the horizontal oscillator section except that the frequency is 15.75 kHz, and loss of horizontal sync results in heavy slanting streaks across the screen. This phenomenon can be witnessed by simply misadjusting the horizontal control on a TV set.

Vertical Deflection Circuit

The vertical deflection system in Fig. 9-25 forms the approximate 60-Hz signal through multivibrator action between transistors Q_{302} and Q_{306}. The required feedback for multivibrator oscillation is provided by R_{308}, R_{320}, vertical hold potentiometer R_{328}, and C_{310}.

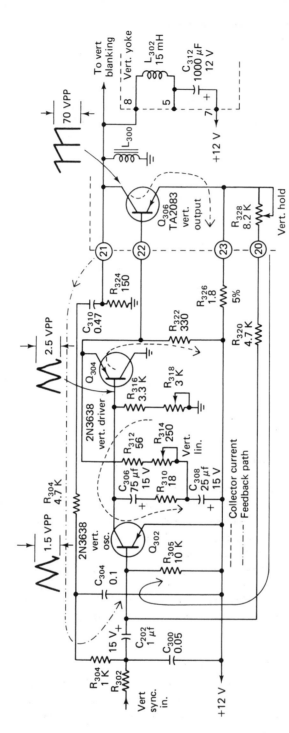

Fig. 9-25. Vertical deflection system.

The vertical sync pulses are detected by the low-pass filter (integrator) made up of R_{302} and C_{300} and coupled into the base of Q_{302} by C_{302}. This synchronizes the multivibrator to precisely 60 Hz, assuming the vertical hold is set to within an approximate correct range. During the vertical scanning period, Q_{302} is cut off, while the sawtooth-forming capacitors C_{306} and C_{308} discharge through the series combination of R_{316} and R_{318}.

This current develops a sawtooth waveform at the base of Q_{304}, which drives the vertical output transistor Q_{306}. Adjustment of R_{318} controls the sawtooth amplitude and subsequently the picture height. It is usually available at the rear of the set to allow periodic adjustment to compensate for component aging.

The vertical output signal is applied to the vertical yoke to magnetically deflect the CRT electron beam vertically downward in a linear fashion. A sawtooth waveform is required for this, and the waveform linearity control R_{314} allows adjustment of this waveform. Incorrect setting may vertically "stretch out" or "squeeze" a portion of the picture. This control is normally available at the rear of the set. The vertical output transistor Q_{306} also has a takeoff point to cause the CRT electron beam to be turned off during the vertical retrace interval. During retrace, Q_{306} is cut off, causing the magnetic field of L_{300} to rapidly collapse, which causes C_{312} to discharge. This returns the CRT electron beam from the bottom to the top of the screen so as to begin the next field.

9-11 HORIZONTAL DEFLECTION AND HIGH VOLTAGE

A typical block diagram for the horizontal deflection and high-voltage systems is shown in Fig. 9-26. The horizontal sync pulses are used to calibrate the horizontal oscillator, which is then amplified to a powerful level by the horizontal output amplifier and then applied to a high-voltage transformer commonly referred to as the *flyback transformer*. Its outputs drive the horizontal yoke windings and provide the required CRT high voltages after rectification. A *damper* function is also provided, which will be explained subsequently. The horizontal system is seen to be a complex one, and because of this and the high voltages and powers involved, it is probably the most failure-prone section of a TV receiver.

As with vertical scanning, a linear sawtooth (current) waveform is required for linear horizontal deflection. If not, distortion in the picture results, as indicated in Fig. 9-27. A horizontal linearity control is usually provided at the rear of the set to correct for these conditions.

Horizontal Deflection Circuit

Figure 9-28 provides a schematic for a typical horizontal system. The horizontal oscillator frequency is held in sync by comparing the horizontal sync pulses to a signal fed back from the horizontal output in the phase detector diodes, D_1 and D_2. Any difference is detected as a phase difference and applied as a dc level to the base of the horizontal oscillator to correct its frequency. Note also the user-

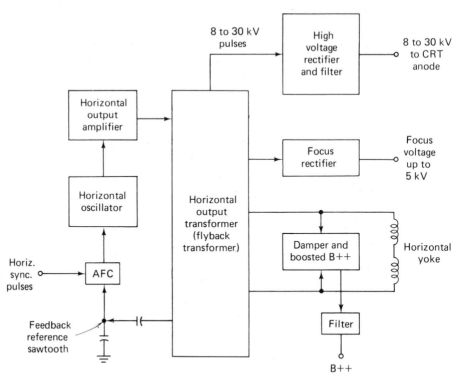

Fig. 9-26. Horizontal deflection block diagram.

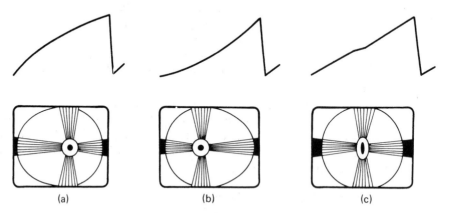

Fig. 9-27. Nonlinear horizontal scanning. (a) Stretching at left and crowding at right. (b) Reverse effect. (c) Crowding at center. (Courtesy of McGraw-Hill, Inc., from Basic Television Principles and Servicing, 4th ed., Grob, Bernard.)

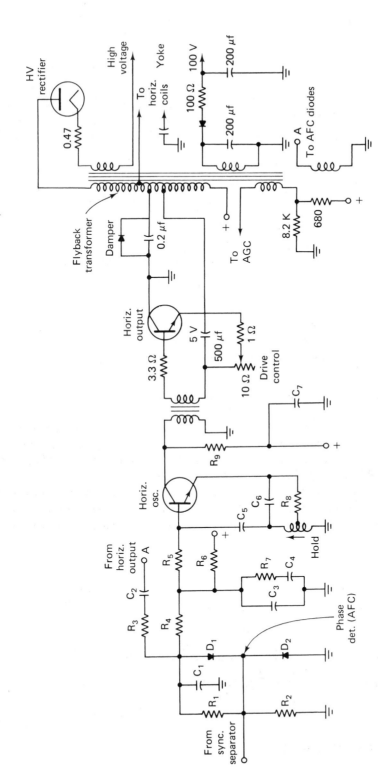

Fig. 9-28. Horizontal system schematic.

controlled variable inductor that adjusts the frequency of oscillation into the range that allows the sync pulses to exercise control. The horizontal oscillator signal is transformer-coupled into the base of the horizontal output transistor for amplification so that the signal has sufficient strength to drive the flyback transformer.

As the sawtooth level builds up on the horizontal amp base, its collector current builds up through the transformer primary and *damper* diode. When the sawtooth level suddenly changes (during retrace), the collector current drops to zero. The magnetic field around the horizontal yoke coils collapses, rapidly inducing a high-amplitude flyback EMF across the transformer secondary. This induces a pulse of current in the secondary of the transformer and a high induced flyback EMF in the primary. The kilovolts of ac thus induced are rectified by the high-voltage rectifier and applied to the CRT as its required dc anode voltage.

During this flyback period, the energy of the horizontal yoke coil's collapsing magnetic field tends to produce damped oscillations that interfere with the start of the next sawtooth waveform. The damper diode serves as a short during this flyback interval so that the unwanted oscillations are very quickly damped.

An auxiliary secondary winding on the flyback transformer provides a stepped *voltage boost* dc level of 100 V for all the circuitry requiring more than the 12 to 20 V dc used elsewhere.

9-12 PRINCIPLES OF COLOR TELEVISION

We have thus far been mainly concerned with black-and-white or *monochrome* television. While color TV presents a much greater degree of sophistication, the reader who has mastered monochrome principles reasonably well can advance to the color set by adding a few more basic ideas.

Our system for color TV was instituted in 1953 and is termed *compatible*. That is, a color transmission can be reproduced in black-and-white shades by a monochrome receiver and a monochrome transmission is reproduced in black and white by a color receiver.

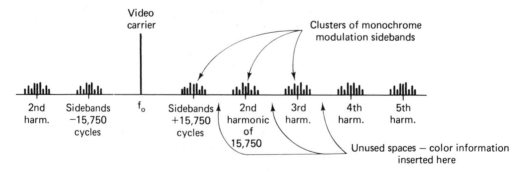

Fig. 9-29. Interleaving process.

To remain compatible, the same total 6-MHz bandwidth must be used, but more information (color) must be transmitted. This problem is overcome by a form of multiplexing, as when FM stereo was added to FM broadcasting. It turns out that the video signal information is clustered at 15.75-kHz intervals throughout its 4-MHz bandwidth. Midway between these 15.75-kHz clusters of information are unused spaces, as indicated in Fig. 9-29. By generating the color information around just the right color subcarrier frequency (3.579545 MHz) it becomes centered in clusters exactly between the black-and-white signals. This is known as *interleaving*.

At the color TV transmitter the scene to be televised is actually scanned by three separate cameras, each camera being sensitive only to one of the three colors of red, blue, and green. Since various combinations of these three colors can be mixed to form any color to which the human eye is sensitive, an electrical representation of a complete color scene is possible. The three color cameras scan the scene in unison, with the red, green, and blue color content separated into three different signals. At the receiver these three separate signals are made to illuminate groups of red, green, and blue phosphor dots (called *triads*), and the original scene is reproduced in color.

After generation these three separate color signals are fed into the transmitter signal processing circuits (*matrix*) and create the Y, or *luminance*, signal and the *chroma*, or color, signals I and Q. The Y signal contains just the right proportion of red, blue, and green such that it creates a normal black-and-white picture. It modulates the video carrier just as does the signal from a single black-and-white camera with a 4-MHz bandwidth. The chroma signals, I and Q, are used to modulate the 3.58-MHz color subcarrier, which then *interleaves* their color information in the gaps left by the luminance Y signal's side bands. This modulation by the I and Q signals is accomplished in a balanced modulator, thus suppressing the 3.58-MHz subcarrier, as it would cause interference at the receiver. The composite transmitted signal is shown in Fig. 9-30.

At the receiver, a monochrome set will simply detect the Y signal and thus present a normal black-and-white rendition of a color picture. The chroma signals (I and Q) cannot be detected in a monochrome set because their 3.58-MHz subcarrier was suppressed and is not present in the received signal. Thus, a color set must have a means to generate and reinject the 3.58-MHz subcarrier to enable detection of the I and Q signals.

Color Receiver Block Diagram

A block diagram of a color receiver from the video detector on is shown in Fig. 9-31. The areas shown are the basic differences between it and a monochrome receiver. After video amplification the Y signal is immediately available. It is given a delay of about 1 μs, as shown, so that it will arrive at the CRT at the same time as the I and Q signals. This is necessary because the I and Q signals undergo considerably more processing, which takes about 1 μs. The chroma signals are ampli-

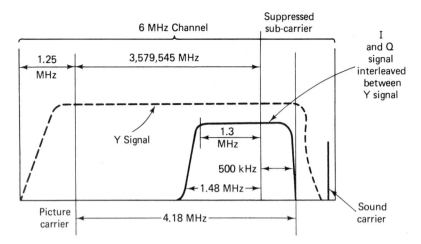

Fig. 9-30. Composite color TV transmission.

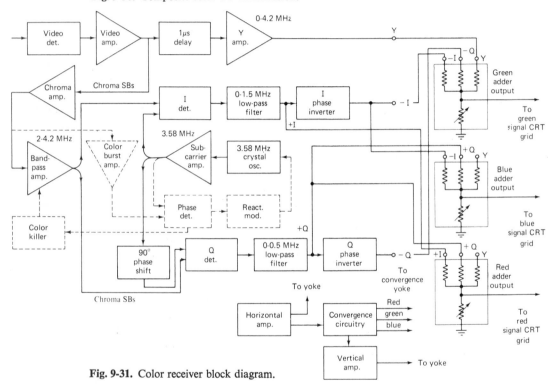

Fig. 9-31. Color receiver block diagram.

fied and then sent into a 2- to 4.2-MHz bandpass amplifier and are then sent into an I detector and a Q detector. These detectors also have inputs from the 3.58-MHz crystal oscillator such that the difference signal in the I detector is the 0- to 1.5 MHz I signal and from the Q detector is the 0- to 0.5-MHz Q signal. Notice in Fig. 9-31 that the 3.58-MHz signal for the Q detector is given a 90° phase shift, which is how the Q signal was generated at the transmitter. It is this phase shift that makes them separable at the receiver.

Once the I and Q signals are detected and passed through their respective low-pass filters, they are given a phase inversion that allows for both + and − chroma signals. This is necessary since

$$\text{green} = -I - Q + Y$$
$$\text{blue} = -I + Q + Y$$
$$\text{red} = +I + Q + Y$$

The I, Q, and Y signals are summed in the three-color adder circuits, with the resistor values providing the proper proportion of each signal. The output of each color adder is then applied to the appropriate CRT grid to control beam intensity. Notice the rheostat in each adder circuit. It allows for the intensity of each color signal to be varied in proportion to the other colors.

The color subcarrier crystal oscillator is not precise enough by itself to allow proper chroma signal detection. This is surprising since crystal oscillators are extremely stable and accurate. An accuracy of 1 part in 10^{12} is necessary to obtain the correct chroma signal. Recall that color transmissions eliminate this carrier from the video signal but do include a sample of it on the back porch of the horizontal blanking pulse, as shown in Fig. 9-32. The color burst amp shown in Fig. 9-31 is receptive to that portion of the overall video signal. Its frequency is compared with the 3.58-MHz crystals in the phase detector, and if they are not precisely equal, the phase detector applies a dc level to vary the reactance of the reactance modulator. It, in turn, causes the crystal's frequency to "pull" in the proper direction to bring it back into precise synchronization with the color burst frequency.

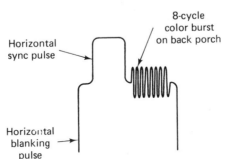

Fig. 9-32. Color burst.

Notice that the phase detector in Fig. 9-31 also has an output that is applied to the *color killer*. The name is very descriptive since a monochrome broadcast has no color burst, and thus the phase detector has a large dc output that the color killer circuit uses to "kill" the 2- to 4.2-MHz bandpass amplifier. The purpose is to prevent any signals out of the chroma circuits during a monochrome broadcast. A defective color killer results in colored noise, called *confetti*, on the screen of a color receiver during a black-and-white transmission. The confetti looks like snow but with larger spots, in color.

The Color CRT and Convergence

Color receiver CRTs are a marvel of engineering precision. As previously mentioned, they are made up of triads of red, blue, and green phosphor dots. The trick is to get the proper electron beam to strike its respective colored phosphor dot. This is accomplished by passing the three beams through a single hole in the *shadow mask*, as shown in Fig. 9-33. The shadow mask prevents the "red" beam from spilling over onto an adjacent blue or green phosphor dot, which would certainly destroy the color rendition. A typical color CRT has over 200,000 holes in the shadow mask and triads of phosphor dots. To make the three beams properly "converge" on their color dot of phosphor throughout the face of the tube requires special modification to the horizontal and vertical deflection systems.

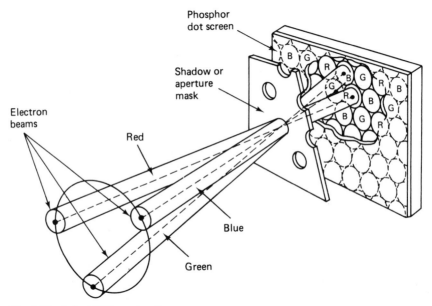

Fig. 9-33. Color CRT construction.

Static convergence refers to proper beam convergence at the center of the CRT's face. This adjustment is made by dc level changes in the horizontal and vertical amplifiers. Convergence away from the center becomes more of a problem and is referred to as *dynamic convergence*. It is necessary because the tube face away from the center is not a perfectly spherical shape (it is more nearly flat), and thus the beams tend to converge in front of the shadow mask away from the tube center. Special dynamic convergence voltages are derived from the horizontal and vertical amplifier signals and are applied to a special color convergence yoke placed around the tube yoke, as shown in Fig. 9-34. The dynamic convergence of a set

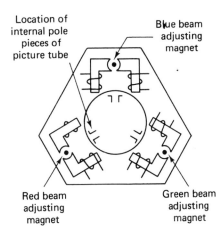

Location of internal pole pieces of picture tube

Blue beam adjusting magnet

Red beam adjusting magnet

Green beam adjusting magnet

Fig. 9-34. Color convergence yoke.

involves the shown magnet adjustment and a number of adjustments on the convergence board, usually 12, that have interaction effects. The process is quite involved and time consuming. The diagram of Fig. 9-31 includes the blocks provided for this process.

9-13 TROUBLESHOOTING

The ability to locate and repair faults in electronic equipment is one of the most important assets of a good technician. Efficient troubleshooting calls for a combination of a logical procedure with a knowledge of the theory of a circuit or system. The theory provided in this chapter combined with your overall electronic knowledge is adequate to allow you to repair the majority of TV failures. If you plan to make a living repairing TVs, however, it will be necessary to make a much more detailed study of TV theory.

The importance of a logical procedure in troubleshooting cannot be overemphasized. The following guide can be applied to the repair of any electronic equipment as well as TVs. As a preliminary check:

1. Verify that the equipment malfunctions as described. If it does not malfunction, the equipment should be operated for a long period of time to determine if the problem is intermittent or induced by a gradual heat buildup.
2. Once the problem has been verified, a visual inspection will often pinpoint it. A thorough cleaning of the equipment should be performed prior to these steps.
   A. With the power off, check for open fuses or tripped circuit breakers, charred components or insulation, loose connections, burned circuit board traces, or any other visual abnormalities.

B. With power applied, *listen* for any unusual sounds, such as vibrating transformers, arcing, and oscillations. Also *smell* around for any components giving off a strong odor due to overheating. Carefully *feel* components suspected of overheating. *Visually* check for dead pilot lights and vacuum tubes that do not light up.

C. With power still applied, lightly tap active devices, components, and controls. Carefully stress leads and connections. Note any change in operation during this process.

If the preceding preliminary checks fail to localize the problem, a more in-depth check is required. Often the symptoms of the equipment will help to isolate the problem. In TV the problems relate to the following readily apparent symptoms:

1. No picture, sound, or raster. This indicates a power supply problem, as none of the subsystems are operational. The *raster* is the illumination of the picture tube by the scanning lines when no received signal is present.

2. Sound normal but no raster. This indicates a problem in the horizontal and/or vertical oscillator stages or high-voltage supplies.

3. Normal sound and raster but defective or no picture. This indicates a problem in the video amplifiers following the sound takeoff.

4. Normal raster but no (or defective) sound and picture. This indicates a defect in the antenna, RF amps, IF amps, or video amps prior to sound takeoff.

A more detailed procedure for TV failure localization is given as Table 9-1. It is adapted from Matthew Mandl, *Modern Television Systems: Theory and Servicing*, Prentice-Hall, Inc., Englewood Cliffs, N.J., 1974, and is used with permission.

Once the problem is localized to a functional stage, the following steps should be executed with the aid of schematics and information provided by the manufacturer:

1. Check static measurements and compare with those supplied by the manufacturer.

2. If step 1 fails to pinpoint the failure, check dynamic (ac) circuit waveforms.

3. When a discrepancy is found, review in your mind all possible causes.

4. Check out each possible cause until the fault is discovered, and make the repair.

5. Make a full operational check to ensure that the equipment is fully repaired, and allow it to operate for at least several hours. The repair can now be considered complete.

TABLE 9-1

Master Index to Common Television Troubles

Symptoms	Probable Cause*	Symptoms	Probable Cause*
No picture—sound normal		Picture normal—sound poor (or missing)	
No picture but raster present	Defect in stages after sound pickoff		Gassy audio amplifier tube or transistor
No picture or raster	Failure of high-voltage power supply		Improper bias in audio section
	Defective picture tube		Shorted cathode or emitter capacitor
	Excess bias on grid of picture tube		Loose speaker cone or off-center voice coil
	Misadjusted or defective ion trap	Hum from speaker	Cathode-filament short in tube-type audio section
	Defective flyback sweep system		Filter capacitor in power supply (may also affect picture)
Bright horizontal bar on picture tube	Failure of vertical sweep Open vertical deflection coils	Motorboating sound from speaker	Open second filter in low-voltage power supply
Heavy slanting streaks across screen	Horizontal sweep out of synchronization (improperly set controls or defective sweep circuit components)	Buzz and noise from speaker	Floating grid in tube-type audio section
			Change of alignment in tuner and IF stages
			High-voltage arcing
Heavy slanting streaks across screen—rolling up or down	Improperly set vertical and horizontal controls		Video carrier containing some FM
	Defects in sync separator, clipper, or sync amplifier circuits		Poor contacts in low-voltage supply feed lines
			Defect in limiter stage
	Defects in both sweep systems		Open large capacitor filter across ratio detector output
	Sync clipping in video stages		Loose laminations, loose windings, or arcing within vertical output transformer
Picture normal—sound poor (or missing)			Defective tubes or parts in tuner and IF stages
No sound from speaker	Defect in IF amplifiers, limiter, detector, or audio amplifiers		Incorrect transistor (or tube) voltages in tuner or IF circuits
	Defective loudspeaker		
Weak sound	Improperly set fine-tuning control		
	Change of some component value in local oscillator, causing drift	Both sound and picture missing	
	Defective transistor, tube, or component in sound section	Raster present but no picture or sound	Defective tuner (RF mixer–oscillator) or a defect in any stage from RF to sound takeoff
Sound distorted	Defective interstage coupling capacitor		

305

TABLE 9-1 (Cont.)

Symptoms	Probable Cause*	Symptoms	Probable Cause*
Both sound and picture missing		*Weak picture and sound*	
	A bad transistor, tube, or part in *both* sound and picture sections of receiver		Bad transistor, tube, or part in *both* audio and video amplifiers
	Loose antenna connection or transmission line		Defect in AGC system
No picture, raster, or sound	Low-voltage power supply shorted or failure of low-voltage system	Weak or absent sound pronounced diagonal bar interference on all stations	Low-line voltage
	Defective ac "on–off" switch		Oscillating audio IF amplifier or an oscillating stage prior to sound takeoff
	Open interlock on chassis		Oscillating RF amplifier stage in tuner
	Blown fuse on receiver or ac mains	*Picture not centered—sound normal*	
	Defective ac outlet or line cord	Picture straight but off-center	Misadjusted centering control or defective controls
	A bad transistor, tube, or part in *both* tuner and high-voltage system		Incorrectly positioned focus magnet ring
Sound and picture missing for some channels, not for others	Defective tuner or improper tracking	Picture tilted, slanting to left or right	Misadjusted yoke (deflection coils)
		Part of picture missing at corner	Misadjusted focus magnet ring
			Both focus magnet rings on ion trap misadjusted
Weak picture and sound			Focus and centering magnets misadjusted
Weak but good-quality sound and picture	Defect in antenna system		Yoke too far back on picture tube neck
	Improperly oriented antenna		Focus magnet ring too far back on picture tube neck
	Change in tuner tracking or IF alignment (may affect quality)		Incorrect angle position of focus magnet
	Defective transistor, tube, or part in tuner or in *both* video and sound sections	*Incorrect picture size—sound normal*	
	Decreased voltage from low-voltage power supply	Picture too wide or too high	Misadjusted or defective width control
	Antenna wrong type or too low		Misadjusted or defective height control
Poor quality and weak picture and sound	Change in local oscillator frequency		Incorrect value of parts in sweep circuits
	Change in fine-tuning control	Excessive picture size in both vertical and horizontal directions	Both size controls misadjusted or defective
	Improper tracking and alignment of tuner and IF stages		"Blooming" caused by decrease in high voltage to picture tube

TABLE 9-1 (Cont.)

Symptoms	Probable Cause*	Symptoms	Probable Cause*
Incorrect picture size—sound normal		*Interference on picture—sound normal*	
Insufficient picture width	Misadjusted or defective width control		Oscillating stage in receiver
	Weak transistor or tube in horizontal discharge or horizontal amplifier circuits	Intermittent streaks across picture	Ignition interference due to poor antenna system, poor signal-to-noise ratio in receiver, or too close proximity to impulse noises
	Defective deflection coil or defective parts in horizontal sweep system		Interference from nearby motors or other electrical devices
Insufficient picture height	Misadjusted or defective height control	Intermittent streaks across picture when set is tapped	Microphonic tube or poor socket contacts
	Weak transistor or tube in vertical discharge or amplifier circuits or incorrect value resistors or capacitors		Poorly soldered part in chassis wiring
			Loose lead-in connection at antenna terminal posts
	Defective deflection coil or vertical output transformer	Ragged bar across picture	Diathermy interference
Picture too small	Misadjusted size controls	"Snow" in picture (salt and pepper effect)	Weak station
	Excessive high voltage		Poor signal-to-noise ratio in receiver
	Defective yoke		Defective transistor or tube in any stage from tuner to picture tube
	Defects or degeneration in both sweep sections		
	Decreased potentials from low-voltage power supply		Poor antenna installation or open circuit in antenna system
	Low-line voltage or overloaded power mains		Improperly tracked tuner or misaligned video IF stages
Keystoned picture with reduced width	Shorted capacitor across one-half of horizontal deflection coil		Improperly adjusted AGC system
	One-half of horizontal deflection coil short-circuited		Defective AGC system
Keystoned picture with reduced height	One-half of vertical deflection coil short-circuited	White vertical bar near center of picture	Misadjusted drive control (left of picture stretched)
	Shorted resistor across one-half of vertical deflection coil		Defective part in horizontal discharge or amplifier circuits
			Defective capacitor or component in damper or voltage boost system
Interference on picture—sound normal		White vertical bars (one or two) near left of picture or raster	Defects in damping circuit
Slanted, tweedy lines across picture	Interference from short-wave station or other heterodyne interference		Defects in voltage boost system
	Local oscillator interference from nearby television receivers		

TABLE 9-1 (Cont.)

Symptoms	Probable Cause*	Symptoms	Probable Cause*
Interference on picture—sound normal		*Interference on picture—sound normal*	
	Defects in horizontal sweep output system	Framing vertical and horizontal bar pattern moving across screen	Interference from upper adjacent channel (picture carrier interference)
Black vertical bars (one or two) near left of picture or raster	Transient oscillations (Barkhausen) in horizontal output amplifier		Mis-adjusted upper adjacent channel trap
	Defective horizontal output tube		Improper antenna orientation
	Improper lead dress in horizontal output circuit		Poor front-to-back antenna ratio
Lacy (moire) effect in picture	Loss of interlace	Double image picture	Interference from reflected signal
	Defective component in integrator circuit	Repeat lines at right of white-to-black or black-to-white objects	"Echo effect" due to high-frequency transients in video amplifier stage
Horizontal bars on screen	Interference from frequency below 15,750 but above 60 (square-wave type of interference)		Overpeaked video amplifier
			Incorrect value peaking coils
Vertical bars on screen	Interference from frequency above 15,750		Change in transistor, tube, or component part values in video amplifiers
Audio bars on screen	Sound interference in picture	Retrace lines showing in picture	Misadjusted or weak ion trap
	Misadjusted sound traps		Improper blanking
	Co-channel interference		Weak signal
Horizontal bar at top of picture	Open damping resistor across vertical coil section		Defective picture tube
Thick black horizontal bar across screen	Cathode heater shorted in picture stages	Hum bar visible only when station is tuned in	Cathode to filament short in local oscillator tube of tuner
	Hum (ac) in video signal	Hum from speaker accompanies appearance of bar	Fine tuning Mis-adjusted
	Improper shielding or lead dress		
Bar with edge pull or ripple	Hum (ac) entering deflection (sweep) systems as well as picture stages	Interruption of vertical detail by horizontal line structure	Interaction between sweep circuits
	Open filter capacitor in low-voltage system (usually noticeable hum also heard from speaker)	Jagged horizontal bar	Arcing in defective lighting fixture or in the 60-Hz ac mains (bar interference may move vertically or may lock in the vertical sweep)
Thin, very closely spaced vertical lines	Defective or misadjusted 4.5-MHz trap in video amplifier stage		
Vertical black bar down center of picture—picture in two sections, back to back	Incorrect phase in horizontal lock sync system	*Poor picture quality—sound normal*	
	Maladjusted phase control	Excessively dark picture	Defect in AGC system
	Defective transistor, tube, or component in horizontal lock system		Defective dc restorer circuit

TABLE 9-1 (Cont.)

Symptoms	Probable Cause*	Symptoms	Probable Cause*
Poor picture quality—sound normal		*Poor picture quality—sound normal*	
	Improperly adjusted AGC control	ture when contrast is advanced	
	Excess signal input (sets without AGC)	Repeat lines following abrupt contrast changes	Overpeaked video IF or video amplifier stages (echo effect)
	Defective contrast control		Improper lead dress in video amplifiers
Dark picture with trailing smears	Poor low-frequency response in video amplifiers		Change in part values in video amplifiers
	Improperly tracked tuner		
	Improperly aligned video IF	*Unsteady and pulling picture—sound normal*	
	Defective decoupler circuit in video amplifier stages	Picture weaving at one side or edge	Magnetic field near picture tube
Negative picture	Defective picture tube or signal overload		Magnetized metal near picture tube (stationary bend in picture)
	Incorrectly set oscillator slug	Picture weaving at both sides (plus excessive contrast)	Excessive signal strength at picture tube grid
	Open peaking coil		Defective or misadjusted AGC
	Defective oscillator (local) (will also affect sound)		Defective contrast setting or control in sets without AGC
	Defective AGC transistor, tube, or circuit		Unstable horizontal lock system
Indistinct picture	Misadjusted focus magnet or focus electrode voltage	Picture weaving at sides, normal contrast	Ripple (ac) in horizontal sweep
	Defective focus control		Misaligned video IF stages causing sync amplitude decrease
	Reflections on transmission line		
	Improper alignment		Poor low-frequency response in video amplifier prior to sync takeoff
	Defective picture tube		
	Defective peaking coils		
	Defective video amplifier stages		Interference present in video signal
	Insufficient high voltage on second anode of picture tube	Picture weaving, poor contrast, and snowy picture	Weak signal
Only a portion of picture area in focus	Improper spacing between focus magnet and yoke		Misaligned tuner or IF stages
	Incorrect voltage in focus electrode circuit		Defective transistor, tube, or part in tuner or IF stages
	Misplaced or weak-focus magnets		Defective circuit prior to sync takeoff
Faded picture with normal contrast setting—improper relation between "tones" of pic-	Poor emission of picture tube		

TABLE 9-1 (Cont.)

Symptoms	Probable Cause*	Symptoms	Probable Cause*
Unsteady and pulling picture—sound normal		*Unsteady and pulling picture—sound normal*	
Picture jitters and has severe streaks	Interference from nearby violet-ray machine, cash register, electric shaver, or other local noise-generating equipment		Defective leveling in sync separators feeding synchroguide or phase detector
Unstable vertical synchronization	Misadjusted vertical hold control	Unstable horizontal synchronization	Misaligned horizontal lock system
	Defective transistor, tube, or part in vertical oscillator stage		Defective diode, transistor, tube, or part in horizontal oscillator and control
	Too critical a setting of AGC control		Defect in any part of horizontal sweep system interacting on horizontal oscillator
	Defective AGC system (horizontal sync will be less affected)		Sync pulse attenuation in stages prior to AFC system
	Defect in sync separator to vertical input circuit		Misadjusted horizontal hold control
Intermittent picture shift to left or right	Incorrectly adjusted horizontal lock system	Both vertical and horizontal sweep unstable	Defective diode, transistor, tube, or part in sync separator or amplifier stages
	Defective horizontal centering control		Sync tip attenuation in video stages because of poor low-frequency response
	Intermittent voltage change		
Intermittent picture shift in vertical plane	Defective vertical centering control		Improperly adjusted or defective AGC system
	Intermittent voltage change in vertical system		Poor signal strength
	Defective vertical oscillator or output tube		Defects in *both* vertical and horizontal sweep systems
Slight weave at top of picture	Critically adjusted horizontal lock system		Defective dc restorer diode or circuit from which signal is taken for sync separator stages
	Improper sync amplitude because of loss at sync takeoff or in sync separator circuits		Incorrect voltage amplitude to sync separator or sweep stages
	Sync clipping in video amplifiers		
Picture weave on strong stations (or when fine tuning set for maximum picture signal)	Defect in AGC system	*Intermittent picture and/or sound*	
	Shorted bias in tuner stages	Intermittent picture—sound normal	Defects or loose connections in IF stages *after* sound takeoff
	Shorted bias in video IF stages		Defective parts or loose connections in detector
	Defective contrast control (in receivers without AGC)		

310

TABLE 9-1 (Cont.)

Symptoms	Probable Cause*	Symptoms	Probable Cause*
Intermittent picture and/or sound		*Picture folds over or crowds—sound normal*	
	or video amplifiers		Defective transistors, tubes, or parts in vertical sweep system
	Loose picture tube socket or loose leads to picture tube socket	Picture information crowded at top	Incorrect vertical linearity
	Defective parts or loose connections in video amplifier, tube socket, or leads to picture tube		Defects in vertical sweep system
Intermittent picture and sound	Defective parts or loose connections in any stage from tuner to picture tube grid	Picture information crowded at bottom	Incorrect vertical linearity
			Defects in vertical sweep system
	Defects in picture tube socket	Picture information crowded at left	Incorrect horizontal linearity
Intermittent sound —picture normal	Defective parts or loose connections from sound takeoff to speaker		Defects in horizontal system from discharge to deflection coils
Intermittent streaks in picture— arcing sound	Intermittent short or arcing in high-voltage power supply	Picture information crowded at right	Incorrect horizontal linearity
	Defective high-voltage rectifier tube or solid-state diode rectifier		Defects in horizontal system from discharge to deflection coils
Intermittent picture, raster, and sound	Intermittent failure of low-voltage power supply system		Incorrectly set or defective drive control (also causes stretching at left)
	Recurrent short across voltage source	Picture stretches out at left but fills mask fully	Misadjusted drive and width controls
Picture folds over or crowds—sound normal		*Color receiver defects†*	
Picture folds over at left	Incorrect retrace versus blanking intervals in horizontal sweep	Tinted black-and-white picture	Purity out of adjustment
			Picture tube requires degaussing
	Defective damping circuit in tube and solid-state units	Some colors more vivid than others	Misadjusted screen or drive controls
Picture folds over at right	Poor low-frequency response in sweep amplifier (horizontal)	Green tones lacking	Defects in green matrix system or G-Y tube or transistor
Excessive foldover	Loss of harmonic components of horizontal sweep waveform resulting in virtual sine-wave sweep	Colors not properly registered	Convergence out of adjustment
		Poor focus	Defective focus rectifier or circuit
		Screen predominantly red	Defective B-Y demodulator transistor or tube
			Defective demodulator circuit
Picture folds over at bottom or top	Leaky coupling capacitor in vertical output		Green and blue drive or screen controls misadjusted

311

TABLE 9-1 (Cont.)

Symptoms	Probable Cause*	Symptoms	Probable Cause*
Color receiver defects†		*Color receiver defects*	
Screen predo-minantly blue	Defective R-Y demodulator transistor or tube		Defective or improperly set color control
	Defective demodulator circuit	Intermittent streaks of color	Arcing in high-voltage system
	Green and red drive or screen controls mis-adjusted	Horizontal color bar interference	Hum (ac) in demodulator or color amplifier circuits
Loss of color	Defects in transistors, tubes, or components of demodulator stages	Color brilliancy levels unstable	Defective shunt regulator tube in high-voltage system
	Color killer control improperly set		Defective components in shunt regulator circuits
	Color killer stage defective		Intermittent tubes or defective transistors in chroma stages
	Bandpass amplifier stage inoperative	Tweed lines caused by sound	Incorrect IF alignment
Loss of color sync	Defective crystal, tube, or component in 3.58-MHz oscillator		Misadjusted sound trap or reject potentiometer
	Defective reactance control circuit	Poor color recep-tion for some stations only	Defective or poor antenna system
	Defective burst amplifier	Colors smeared	Loss of Y signal
Faded colors	Improperly adjusted bias control on picture tube		Defective video amplifier system
	Defective picture tube	Moving color bars against black-white background	Loss of color sync
	Improperly adjusted screen and drive controls		Defective burst gate or reactance control
	Weak demodulator transis-tors or tubes	Poor flesh tone colors	Incorrect tint control setting
	Tuner and IF stages require alignment	Flesh tones vary with position on tube face	Purity adjustments incorrect
	Defects in bandpass ampli-fier		

*Probable cause of trouble is, in most instances, based on the assumption the receiver had been working satis-factorily and suddenly developed symptoms detailed. Naturally, if alignment or tracking has been tampered with, sound or picture could be lost because of off-resonance conditions in tuned circuits.

†Defects of width, height, linearity, and the like in a color receiver are caused by the same factors as in black-and-white receivers and are covered in previous sections.

A true measure of a technician's troubleshooting ability is the time it takes to analyze and complete a repair. The major causes of excessive repair time are:

1. A disorganized approach. A logical approach should be used, and it will naturally be improved and refined with experience.

2. Attempting repair without adequate theory or manufacturer's service information. If a preliminary inspection fails to pinpoint the problem in this case, it is best to stop and obtain the necessary information.

3. Failure of the technician to realize that he is "going in circles." When you jump from one check to another time after time, it may be best to try it again tomorrow with a fresh mind.

QUESTIONS AND PROBLEMS

The first number indicates the chapter, the second is sequence, and the third is chapter section. As an example, 1-6-2 indicates chapter 1, the sixth proplem, and from section 2 of the chapter.

CHAPTER 1

1-1-1 Define modulation.

*1-2-1 What is meant by carrier frequency? (3.520)

1-3-1 Describe the two reasons that modulation is used for communication transmissions.

1-4-1 List the three parameters of a high-frequency carrier that may be varied by a low-frequency intelligence signal.

*1-5-2 What are the frequency ranges included in the following frequency subdivisions: MF (medium frequency), HF (high frequency), VHF (very high frequency), UHF (ultra high frequency), and SHF (super high frequency)? (3.522)

1-6-2 List the two basic limitations on the performance of a communication system.

*Indicates a question from the FCC study guide. The number following the question indicates the element it is taken from and the question number. As an example, (3.520) indicates that the question is from element 3 and is question 520.

1-7-3 Define electrical noise and explain why it is so troublesome to a communication receiver.

1-8-3 Explain the difference between external and internal noise.

1-9-3 List and briefly explain the various types of external noise.

1-10-3 Provide two other names for Johnson noise and calculate the noise voltage output of a 1-MΩ resistor at 27°C over a 1-MHz frequency range.

1-11-3 Explain the meaning of the term "low-noise resistor."

1-12-3 Calculate the noise power and voltage outputs for the following:
(a) A 1-kΩ resistor at 27°C over a 100-kHz bandwidth.
(b) A 100-kΩ resistor at 27°C over a 1-kHz bandwidth.

1-13-3 Describe the relationship between shot noise and equivalent noise resistance.

1-14-4 Calculate the S/N ratio for a receiver output of 4 V signal and 0.48 V noise, both as a ratio and in decibel form.

1-15-4 The receiver in problem 1-14-4 has a S/N ratio of 110 at its input. Calculate the receiver's noise figure.

1-16-4 A single-stage amplifier has an input impedance of 75 Ω and is driven by a 75-Ω source, has a 200-kHz bandwidth, and a voltage gain of 100. Assume the external noise and shot noise to be negligible and that a 1-mV signal is applied to the amplifier's input. Calculate the output noise voltage if the amplifier has a 5-dB NF.

1-17-5 Describe the basic differences in noise performance between BJT, JFET, and MOSFET transistors.

1-18-6 Define information theory.

1-19-6 State Hartley's law and explain its significance.

*1-20-7 What is meant by a "harmonic"? (3.232)

*1-21-7 What is the seventh harmonic of 360 kHz? (3.251)

1-22-7 Why does transmission of a 2-kHz square wave require greater bandwidth than for a 2-kHz sine wave?

1-23-7 Draw time *and* frequency-domain sketches for a 2-kHz square wave.

1-24-8 Explain the function of Fourier analysis.

1-25-8 A 2-kHz square wave is transmitted on a 10-kHz bandwidth channel. Sketch the resulting signal and explain why the distortion occurs.

CHAPTER 2

2-1-1 Explain why the *linear* combination of a low-frequency intelligence signal and high-frequency carrier signal is *not* effective as a radio transmission.

2-2-1 A 1500-kHz carrier and 2-kHz intelligence signal are combined in a *nonlinear* device. List *all* the frequency components produced.

*2-3-1   If a 1500-kHz radio wave is modulated by a 2-kHz sine-wave tone, what frequencies are contained in the modulated wave (the actual AM signal)? (3.294)

*2-4-1   If a carrier is amplitude-modulated, what causes the side-band frequencies? (S3.139)

*2-5-1   What determines the bandwidth of emission for an AM transmission? (S3.140)

2-6-1   Describe the significance of the upper and lower envelope of an AM wave-form.

2-7-1   Explain the difference between a side band and a side frequency.

*2-8-2   Draw a diagram of a carrier-wave envelope when modulated 50% by a sinusoidal wave. Indicate on the diagram the dimensions from which the percentage of modulation is determined. (3.328)

*2-9-2   What are some of the possible results of overmodulation? (3.343)

2-10-2   An unmodulated carrier is 300 V p-p. Calculate %m when its maximum p-p value reaches 400 V, 500 V, and 600 V.

2-11-3   A 100-V carrier is modulated by a 1-kHz sine wave. Determine the side-frequency amplitudes when $m = 0.75$.

2-12-3   Calculate the carrier and side-band power if the total transmitted power is 500 W in problem 2-11-3.

2-13-3   The antenna current of an AM transmitter is 6.2 A when unmodulated and rises to 6.7 A when modulated. Calculate %m.

*2-14-3   Why is a high percentage of modulation desirable? (3.342)

*2-15-3   During 100% modulation, what percentage of the average output power is in the side bands? (4.43)

2-16-4   Describe two possible ways that a transistor can be used to generate an AM signal.

*2-17-4   What is meant by "low-level" modulation? (3.336)

*2-18-4   What is meant by "high-level" modulation? (3.334)

2-19-4   Explain the relative merits of high- and low-level modulation schemes.

*2-20-4   Why must some radio-frequency amplifiers be neutralized? (3.279)

2-21-4   Draw a schematic of a class C transistor modulator and explain its operation.

*2-22-4   What is the principal advantage of a class C amplifier? (4.188)

*2-23-4   What is the function of a quartz crystal in a radio transmitter? (3.419)

*2-24-5   Draw a block diagram of an AM transmitter. (53.151)

*2-25-5   What is the purpose of a buffer amplifier stage in a transmitter? (3.298)

2-26-5   Describe the means by which the transmitter shown in Fig. 2-15 is modulated.

*2-27-5   Draw a simple schematic diagram showing a method of coupling the radio-frequency output of the final power-amplifier stage of a transmitter to an antenna. (3.79)

2-28-5   Describe the functions of an antenna coupler.

*2-29-5   A ship radiotelephone transmitter operates on 2738 kHz. At a certain point distant from the transmitter, the 2738-kHz signal has a measured field of 147 mV/m. The second harmonic field at the same point is measured as 405 μV/m. To the nearest whole unit in decibels, how much has the harmonic emission been attenuated below the 2738-kHz fundamental? (3.534)

2-30-5   What is meant by the term "tune-up" procedure?

2-31-6   List the advantages of using a monolithic IC transmitter chip and some possible disadvantages.

2-32-6   List the required circuit changes to change the transmission frequency of the IC transmitter shown in Fig. 2-19(b).

*2-33-7   What is the effect of 10-kHz modulation of a standard broadcast station on adjacent channel reception? (4.164)

2-34-7   Explain why tubes are still used in standard-broadcast-band transmitters.

2-35-7   List some special features found on standard-broadcast-band transmitters that are not normally included on other transmitters.

*2-36-7   What is the purpose of a "dummy antenna"? (3.380)

*2-37-8   Draw a sample sketch of the trapezoidal pattern on a cathode-ray oscilloscope screen indicating low percentage modulation without distortion. (4.42)

2-38-8   Explain the advantages of the trapezoidal display over a standard oscilloscope display of AM signals.

2-39-8   Compare the display of an oscilloscope to that of a spectrum analyzer.

2-40-8   A spectrum analyzer display shows that a signal is made up of three components only: 960 kHz at 1 V, 962 kHz at 1/2 V, 958 kHz at 1/2 V. What is the signal and how was it generated?

2-41-8   Define the meaning of the term "spur."

CHAPTER 3

*3-1-1   Draw a diagram of a tuned radio-frequency-type radio receiver. (3.514)

*3-2-1   Explain: sensitivity of a receiver; selectivity of a receiver. Why are these important quantities? In what typical units are they usually expressed? (S3.161)

3-3-1   List the two factors that determine a receiver's sensitivity. Which one is most important and why?

3-4-1   Explain why a receiver can be overly selective.

*3-5-1   In a parallel circuit composed of an inductance of 150 μH and a capacitance of 160 $\mu\mu$F, what is the resonant frequency? (4.020)

3-6-1   Express the following ratios in decibel form:
(a) Voltage ratio of 210.
(b) Voltage ratio of 110,000.
(c) Voltage ratio of 0.4.
(d) Voltage ratio of 0.0072.
(e) Power ratio of 173.
(f) Power ratio of 0.025.

*3-7-1   What is the Q of a circuit? How is it affected by the circuit resistance? How does the Q of a circuit affect bandwidth? (S3.34)

3-8-1   A TRF receiver is to be tuned over the 550- to 1550-kHz range with a 25-μH inductor. Calculate the required range of necessary capacitance. Determine the tuned circuit's necessary Q if a 10-kHz bandwidth is desired at 1000 kHz. Calculate the receiver's selectivity at 550 kHz and 1550 kHz.

3-9-2   Why does passing an AM signal through a nonlinear device allow recovery of the original (low-frequency) intelligence signal when the AM signal contains only high frequencies?

*3-10-2   Explain the operation of a diode type of detector. (3.130)

3-11-2   Describe the advantages and disadvantages of a diode detector.

3-12-2   Explain how diagonal clipping occurs in a diode detector.

3-13-2   Provide the advantages of a synchronous detector compared to a diode detector.

*3-14-3   Draw a block diagram of a superheterodyne AM receiver. Assume an incident signal and explain briefly what occurs in each stage. (S3.155)

*3-15-3   What type of radio receivers contains intermediate-frequency transformers? (3.264)

3-16-3   Explain how the superheterodyne receiver allows for constant selectivity over an entire band of received frequencies.

3-17-3   The AM signal into a mixer is a 1.1-MHz carrier that was modulated by a 2-kHz sine wave. The local oscillator is at 1.555 MHz. List all mixer output components and indicate those "accepted" by the IF amplifier stage.

*3-18-3   Explain the purpose and operation of the first detector in a superheterodyne receiver. (3.273)

3-19-3   Explain how the variable-tuned circuits in a superheterodyne receiver are adjusted with a single control.

3-20-4   Provide an adjustment procedure whereby adequate tracking characteristics are obtained in a superheterodyne receiver.

3-21-4   Draw a schematic that illustrates "electronic" tuning using a varactor diode.

*3-22-5   If a superheterodyne receiver is tuned to a desired signal at 1000 kHz, and its conversion (local) oscillator is operating at 1300 kHz, what would be the frequency of an incoming signal that would possibly cause "image" reception? (3.354)

*3-23-5   Explain the relation between the signal frequency, the oscillator frequency, and the image frequency in a superheterodyne receiver. (3.464)

3-24-5   Show why image frequency rejection is not a major problem for the standard AM broadcast band.

*3-25-5   What are the advantages to be obtained from adding a tuned radio-frequency amplifier stage ahead of the first detector (converter) stage of a superheterodyne receiver? (3.351)

*3-26-5   If a tube in the only radio-frequency stage of your receiver burned out, how could temporary repairs or modifications be made to permit operation of the receiver if no spare tube is available? (3.355)

3-27-5   What advantages do dual-gate MOSFETs have over BJTs for use as RF amplifiers?

*3-28-5   What is the "mixer" in a superheterodyne receiver? (6.539)

3-29-5   Describe the advantage of an autodyne mixer over a standard mixer.

3-30-5   Why is the bulk of a receiver's gain and selectivity obtained in the IF amplifier stages?

3-31-6   Describe the difficulties in listening to a receiver without AGC.

*3-32-6   How is "automatic volume control" accomplished in a radio receiver? (3.353)

3-33-6   Explain how the ac gain of a transistor can be controlled by a dc AGC level.

3-34-6   The specs in Fig. 3-5 for the ZN414 LIC TRF receiver indicate an AGC range of 20 dB. What specifically does this mean?

3-35-7   Describe the function of auxiliary AGC.

3-36-7   What is the limiting function with respect to manufacturing a complete superheterodyne receiver on a LIC chip?

CHAPTER 4

4-1-1   Explain why two components of an AM signal (carrier and one side band) may be eliminated and still result in a usable transmission.

4-2-1   An AM transmission of 1000 W is fully modulated. Calculate the power transmitted if it were transmitted as a SSB signal.

4-3-1   The bandwidth of a SSB signal is half that of the corresponding AM signal. Explain why that results in about a 3-dB "gain" at the receiver.

4-4-1   Provide detail on the differences between SSB, SSSC, and ISB transmissions.

*4-5-1   Explain the principles involved in a single-side-band suppressed carrier (SSSC) emission. How does its bandwidth of emission and required power compare with that of full carrier and side bands? (S3.152)

4-6-1 List and explain the advantages of SSB over conventional AM transmissions. Are there any disadvantages?

4-7-2 What typically are the inputs and outputs for a balanced modulator?

4-8-2 Draw a schematic of a balanced modulator using two JFETs, and briefly explain its operation.

4-9-2 What disadvantage might a balanced-ring modulator have, compared to the circuit used in problem 4-8-2?

4-10-2 Referring to the specifications for the SL640 LIC balanced modulator in Fig. 4-5, calculate its typical dc power dissipation and output noise given a 1 mV rms total noise input. Calculate the typical carrier output power given a desired side-band output of 10 μW.

4-11-3 Calculate a filter's required Q to convert DSB to SSB, given that the two side bands are separated by 200 Hz. The suppressed carrier is 29 MHz. Explain how this required Q could be greatly reduced.

*4-12-3 What is a "low-pass" filter? A "high-pass" filter? (4.096)

*4-13-3 Draw a diagram of a simple low-pass filter. (4.097)

*4-14-3 In general, why are filters used? Why are "bandstop," "high-pass," and "low-pass" filters used? Draw schematic diagrams of the most commonly used filters. (S3.036)

*4-15-3 Draw the approximate equivalent circuit of a quartz crystal. (S4.042)

4-16-3 What are the undesired effects of the crystal holder capacitance in a crystal filter, and how are they overcome?

*4-17-3 What crystalline substance is widely used in crystal oscillators (and filters)? (3.248)

*4-18-3 What does the expression "positive temperature coefficient" mean as applied to a quartz crystal? (3.414)

*4-19-3 What are the principal advantages of crystal control over tuned-circuit oscillators (or filters)? (3.423)

4-20-3 Explain the operation and use of mechanical filters.

*4-21-4 Draw a block diagram of a SSSC transmitter (filter type) with a 20-kHz oscillator and emission frequencies in the range of 6 MHz. Explain the function of each stage. (S3.153)

4-22-4 Determine the carrier frequency for the transmitter shown in Fig. 4-12. (It is *not* 3 MHz.)

4-23-4 Briefly explain the principals involved in the ISB transmitter illustrated in Fig. 4-14.

4-24-4 Explain the function of a hybrid coil.

4-25-5 List the advantages of the phase-versus-filter method of SSB generation. Why isn't the phase method more popular than the filter method?

4-26-5 Mathematically show how a DSB signal, $\cos \omega_i t \cos \omega_c t$, can be manipulated to provide SSB.

4-27-5 Explain the operation of the phase shift SSB generator illustrated in Fig. 4-15. Why is the carrier 90° phase shift not a problem while that for the audio signal is?

4-28-6 List the components of an AM signal at 1 MHz when modulated by a 1-kHz sine wave. What are the component(s) if it were converted to an USB transmission? If the carrier is redundant, why must it be "reinserted" at the receiver?

4-29-6 Explain why the BFO in a SSB demodulator has such stringent accuracy requirements.

*4-30-6 Explain, briefly, how a SSSC emission is detected. (S3.154)

4-31-6 If, in an emergency, you had to use an AM receiver to receive a SSB broadcast, what modifications to the receiver would be appropriate?

4-32-7 List the differences required of an ISB receiver compared to an SSB receiver.

4-33-7 What is meant by a "channelized" receiver and why have they been generally replaced by "synthesized" receivers?

4-34-7 Design a frequency synthesizer, in block diagram form, capable of 0-Hz to 9.99-MHz outputs in 10-kHz increments.

4-35-7 List some applications for frequency synthesizers other than in radio.

4-36-7 Design a frequency synthesizer in block diagram form capable of 0-Hz to 99.999-MHz outputs in 1-kHz increments.

4-37-7 From the specifications for the RF-505A receiver in Fig. 4-23, determine its sensitivity for AM and SSB. Explain why it is much more sensitive for SSB than AM when they both have the same first stage of amplification.

CHAPTER 5

5-1-1 Define angle modulation and list its subcategories.

*5-2-1 What is the difference between frequency and phase modulation? (S4.061)

5-3-1 Even though PM is not actually transmitted, provide two reasons that make it important in the study of FM.

5-4-2 Describe the effect of an intelligence signal's amplitude and frequency when it frequency-modulates a carrier.

5-5-2 Explain how a condenser microphone can very easily be used to generate FM.

5-6-2 In an FM transmitter, the output is changing between 90.001 MHz and 89.999 MHz 1000 times per second. The modulating signal amplitude is 3 V. Determine the carrier frequency and modulating signal frequency. If the output deviation changes to between 90.0015 MHz and 89.9985 MHz, calculate the modulating signal amplitude.

*5-7-2 What determines the rate of frequency swing of an FM broadcast transmitter? (4.226)

5-8-3 Define the term modulation index (m_f) as applied to an FM system.

*5-9-3 What characteristic(s) of an audio tone determines the percentage of modulation of an FM broadcast transmitter? (4.225)

5-10-3 Explain the difference between modulation index for PM versus FM. How can a modulating signal be modified so that allowing it to phase-modulate a carrier results in FM?

5-11-3 Explain what happens to the carrier in FM as m_f goes from 0 to 15.

5-12-3 Calculate the bandwidth of an FM system when the maximum deviation (δ) is 15 kHz and $f_m = 3$ kHz. Repeat for $f_m = 2$ kHz and 4 kHz.

*5-13-3 How wide is an FM broadcast channel? (4.243)

5-14-3 Explain the purpose of the "guard" bands for broadcast FM.

*5-15-3 What frequency swing is defined as 100% modulation for an FM broadcast station? (4.244)

*5-16-3 What is the meaning of the term "center frequency" in reference to FM broadcasting? (4.246)

*5-17-3 What is the meaning of the term "frequency swing" in reference to FM broadcast stations? (4.250)

*5-18-3 What is the frequency swing of an FM broadcast transmitter when modulated 60%? (4.219)

*5-19-3 An FM broadcast transmitter is modulated 40% by a 5-kHz test tone. When the percentage of modulation is doubled, what is the frequency swing of the transmitter? (4.228)

*5-20-3 An FM broadcast transmitter is modulated 50% by a 7-kHz test tone. When the frequency of the test tone is changed to 5 kHz and the percentage of modulation is unchanged, what is the transmitter frequency swing? (4.220)

*5-21-3 If the output current of an FM broadcast transmitter is 8.5 A without modulation, what is the output current when the percentage of modulation is 90%? (4.235)

5-22-3 An FM transmitter puts out 1 kW of power. Determine the power in the carrier and all significant side bands when $m_f = 2$. Verify that their sum is 1 kW.

*5-23-3 In an FM radio communication system, what is the meaning of modulation index? Of deviation ratio? What values of deviation ratio are used in an FM radio communication system? (3.494)

*5-24-4 What types of radio receivers do not respond to static interference? (3.247)

*5-25-4 What is the purpose of a limiter stage in an FM broadcast receiver? (4.233)

5-26-4 Explain why the limiter does *not* eliminate all noise effects in an FM system.

5-27-4 Calculate the amount of frequency deviation caused by a limited noise spike that still causes an undesired phase shift of 35° when f_m is 5 kHz.

5-28-4 In a broadcast FM system, the input S/N = 4. Calculate the worst-case S/N at the output if the receiver's internal noise effect is negligible.

5-29-4 Explain why narrow-band FM systems have poorer noise performance than wide-band systems.

5-30-4 Explain the "capture effect" in FM and include the link between it and FM's inherent noise reduction capability.

*5-31-4 Why is narrow-band FM rather than wide-band FM used in radio communication systems? (3.495)

*5-32-4 What is the purpose of preemphasis in an FM broadcast transmitter? Of deemphasis in an FM receiver? Draw a circuit diagram of a method of obtaining preemphasis. (S4.063)

5-33-4 Explain the difference between the Dolby noise reduction system and the standard preemphasis/deemphasis system.

*5-34-4 Discuss the following in reference to frequency modulation.
(a) The production of side bands.
(b) The relationship between the number of side bands and the modulating frequency.
(c) The relationship between the number of side bands and the amplitude of the modulating voltage.
(d) The relationship between percent modulation and the number of side bands.
(e) The relationship between modulation index or deviation ratio and the number of side bands.
(f) The relationship between the spacing of the side bands and the modulating frequency.
(g) The relationship between the number of side bands and the bandwidth of emissions.
(h) The criteria for determining bandwidth of emission.
(i) Reasons for preemphasis. (S3.163)

5-35-5 Draw a schematic diagram of a varactor diode FM generator and explain its operation.

*5-36-5 Draw a schematic diagram of a frequency-modulated oscillator using a reactance modulator. Explain its principle of operation. (S3.162)

5-37-5 Using the specifications in Fig. 5-12, draw a schematic of an FM generator using the SE/NE 566 LIC function generator VCO. The center frequency is to be 500 kHz and the output is to be a sine wave. Show all component values. How much center frequency drift can be expected from a temperature rise of 50°C?

5-38-5 Explain the principles of a Crosby-type modulator.

*5-39-5   How is good stability of a reactance modulator achieved? (S3.164)

*5-40-5   If an FM transmitter employs one doubler, one tripler, and one quadrupler, what is the carrier frequency swing when the oscillator frequency swing is 2 kHz? (4.229)

5-41-5   Draw a block diagram of a broadcast band Crosby-type FM transmitter operating at 100 MHz and label all frequencies in the diagram.

*5-42-6   Draw a block diagram of an Armstrong-type FM broadcast transmitter complete from the microphone input to the antenna output. State the purpose of each stage and explain briefly the overall operation of the transmitter. (S4.089)

5-43-6   Explain the difference to the side bands when passing an FM signal through a mixer as compared to a multiplier.

*5-44-7   Draw a block diagram of a stereo multiplex FM broadcast transmitter complete from the microphone inputs to the antenna output. State the purpose of each stage and explain briefly the overall operation of the transmitter. (S4.089)

5-45-7   Explain how stereo FM is able to effectively transmit twice the information of a standard FM broadcast while still using the same bandwidth.

*5-46-8   What are the merits of an FM communication system compared to an AM system? (3.489)

*5-47-8   Why is FM undesirable in the standard AM broadcast band? (4.046)

CHAPTER 6

*6-1-1   What is the purpose of a discriminator in an FM broadcast receiver? (4.217)

6-2-1   Explain why the automatic frequency control (AFC) function is not necessary in today's FM receivers.

*6-3-1   Draw a block diagram of a superheterodyne receiver designed for reception of FM signals. (6.535)

6-4-2   How does the noise reduction capability of a communication system affect its ultimate sensitivity rating?

6-5-2   Explain the desirability of an RF amplifier stage in FM receivers as compared to AM receivers. Why is this not generally true at frequencies over 1 GHz?

6-6-2   Define the meaning of local oscillator reradiation and explain how an RF stage helps to prevent it.

6-7-2   Why are FETs preferred over BJTs or tubes as the active elements for RF amplifiers?

6-8-2   List two advantages of using a dual-gate MOSFET over a JFET in RF amplifiers.

6-9-2   Explain the need for the radio-frequency choke (RFC) in the RF amplifier shown in Fig. 6-2.

*6-10-3   What is the purpose of a limiter stage in an FM broadcast receiver? (4.233)

*6-11-3   Draw a diagram of a limiter stage in an FM broadcast receiver. (4.238)

6-12-3   Explain fully the circuit operation of the limiter shown in Fig. 6-3.

6-13-3   Explain why a limiter minimizes or eliminates the need for the AGC function.

6-14-3   What is the relationship between limiting, sensitivity, and quieting for an FM receiver?

6-15-3   An FM receiver provides 100 dB of voltage gain prior to the limiter. Calculate the receiver's sensitivity if the limiter's quieting voltage is 300 mV.

6-16-4   Draw a schematic of an FM slope detector and explain its operation. Why is this method not often used in practice?

6-17-4   Draw a schematic of a Foster–Seely discriminator and provide a step-by-step explanation of what happens when the input frequency is below the carrier frequency. Include a phase diagram in your explanation.

*6-18-4   Draw a diagram of an FM broadcast receiver detector circuit. (4.236)

*6-19-4   Draw a diagram of a ratio detector and explain its operation. (S3.178)

6-20-4   Explain the relative merits of the Foster–Seely and ratio detector circuits.

*6-21-4   Draw a schematic diagram of each of the following stages of a superheterodyne FM receiver. Explain the principles of operation. Label adjacent stages.
(a) Mixer with injected oscillator frequency.
(b) IF amplifier.
(c) Limiter.
(d) Discriminator. (S3.177)

*6-22-4   Draw a block diagram of a superheterodyne receiver designed for reception of FM signals. (6.535)

6-23-5   Draw a block diagram of a phase-locked loop (PLL) and briefly explain its operation.

6-24-5   Explain in detail how a PLL is used as an FM demodulator.

6-25-5   List the three possible states of operation for a PLL and explain each one.

6-26-5   A PLL's VCO free-runs at 7 MHz. The VCO does not change frequency until the input is within 20 kHz of 7 MHz. After that condition the VCO follows the input to ±150 kHz of 7 MHz before the VCO starts to free-run again. Determine the PLL's lock-and-capture range.

6-27-5   Draw a schematic of the 560 PLL of Fig. 6-14 used as an FM demodulator. Pick C_o for operation at 10.7 MHz and C_D to provide a 75-μs deemphasis time constant. What is the typically required input signal and how much output is to be expected?

6-28-6   Draw a block diagram for an FM stereo demodulator. Explain in detail the function of the *AM* demodulator and the matrix network. Make a *circuitry* addition that simply energizes a light to indicate reception of a stereo station.

6-29-6   Explain how separate left and right channels are obtained from the L + R and L − R signals.

*6-30-6 What is SCA? What are some possible uses of SCA? (S4.115)

6-31-6 Determine the maximum reproduced audio signal frequency in an SCA system. Why does SCA cause less FM carrier deviation and is thus less noise resistant than standard FM? (*Hint:* Refer to Fig. 6-17.)

6-32-6 Explain the principle of operation for the CA 3090 stereo decoder.

6-33-7 The receiver front end in Fig. 6-20 is rated to have noise below the signal by 30 dB in the output with a 1.75-μV input. Calculate its output S/N ratio with a 1.75-μV input signal. (*Answer:* 31.62:1.)

6-34-7 The LIC dual audio amplifiers in Fig. 6-20 are rated to provide 70 dB of channel separation. If the left channel has 1 W of output power, calculate the wattage of the right channel that is included. (*Answer:* 0.1 μW.)

CHAPTER 7

7-1-1 Explain the difference between an FM stereo receiver and a communication transceiver.

7-2-1 Draw a block diagram for a double-conversion receiver when tuned to a 27-MHz broadcast using a 10.7-MHz first IF and 1-MHz second IF. List all pertinent frequencies for each block. Explain the superior image frequency characteristics as compared to a single-conversion receiver with a 1-MHz IF and provide the image frequency in both cases.

7-3-1 Describe the process of up-conversion. Explain its advantages and disadvantages, compared to double conversion.

7-4-1 Draw block diagrams and label pertinent frequencies for a double-conversion *and* up-conversion system for receiving a 40-MHz signal. Discuss the economic merits of each system and the effectiveness of image frequency rejection.

7-5-1 Discuss the advantages of delayed AGC over normal AGC and explain how it may be attained.

7-6-1 Explain the function of auxiliary AGC and a means of providing it.

7-7-1 List two methods of obtaining band spread on a receiver and explain its function.

7-8-1 Explain the need for variable sensitivity and show with a schematic how it could be provided.

*7-9-1 What would be the advantages and disadvantages of utilizing a bandpass switch on a receiver? (S3.160)

7-10-1 What is the need for a noise limiter circuit? Explain the circuit operation of the noise limiter shown in Fig. 7-8.

7-11-1 List some possible applications for "metering" on a communications transceiver.

*7-12-1 What is the purpose of a squelch circuit in a radio communication receiver? (3.496)

7-13-1 List two other names for a squelch circuit. Provide a schematic of a squelch circuit and explain its operation.

7-14-2 Describe the operation of an automatic noise limiter (ANL).

7-15-2 Explain the need for, and function of, "delta tune" in a CB transceiver.

7-16-4 In general terms, explain the facsimile process.

7-17-4 A standard TV broadcast transmits a full picture in $\frac{1}{30}$ second. Why does a typical facsimile transmission require several minutes?

7-18-5 In general terms, explain the operation of the mobile radio telephone system shown in Fig. 7-13.

7-19-6 Provide the following specifications for the RF-280 communication transceiver.
(a) Transmitter output power.
(b) FM deviation.
(c) Receiver sensitivity.
(d) Image frequency rejection.
(e) Receiver antenna radiation.

7-20-6 Trace the receive signal flow of the RF-280 transceiver shown in Fig. 7-17. Briefly explain the function of each block in the process.

7-21-6 Trace the transmit signal flow of the RF-280 transceiver shown in Fig. 7-17. Briefly explain the function of each block in the process.

CHAPTER 8

8-1-1 In what ways is coded voice transmission advantageous over direct transmission? What are some possible disadvantages?

8-2-1 Provide a definition for "coding" and "bit."

8-3-1 Determine the number of bits required to encode a system of 50 equiprobable events with a binary code. Calculate the efficiency of this code. Calculate the efficiency of a decimal code to accomplish the same goal.

8-4-1 Explain the noise immunity advantages of the binary code over any other code.

8-5-1 Define "redundancy" and explain its usefulness with respect to code transmissions.

8-6-1 Provide the Baudot code for "REDSKINS 23 FORTYNINERS 20."

8-7-1 Explain the major disadvantage of the Morse code for machine-controlled communication systems.

8-8-2 Define what is meant by a continuous-wave transmission. In what way is this an inappropriate name? Explain the role that a keying filter plays in a CW transmission.

8-9-2   Explain the AGC difficulties encountered in the reception of CW. What is two-tone modulation and how does it remedy the receiver AGC problem of CW?

8-10-2   Calculate all possible transmitted frequencies for a two-tone modulation system using a 21-MHz carrier with 300-Hz and 470-Hz modulating signals to represent mark and space. Calculate the channel bandwidth required.

8-11-2   What is a frequency-shift keying system? Describe several methods of generating FSK.

*8-12-2   Explain briefly the principles involved in frequency-shift keying (FSK). How is this signal detected? (S3.175)

8-13-2   Explain the functions of a teleprinter. Include in this explanation the meaning of a "baud" and how it relates to the standard transmission rate of 60 words per minute.

8-14-2   Describe the advantages of using a perforated-tape unit in conjunction with a teleprinter.

8-15-3   Describe the key distinction between pulse modulation and amplitude or frequency modulation.

8-16-3   List two advantages of pulse modulation and explain their significance.

8-17-3   A signal that varies from 20 Hz to 5 kHz is to be processed via a pulse modulation scheme. Determine the minimum sampling rate that will still allow adequate reproduction at the receiver. Calculate the number of different time-division-multiplexed signals that could be transmitted if each sample takes 10 μs.

8-18-3   With a sketch similar to Fig. 8-11, explain the basics of PAM, PDM, and PPM.

8-19-3   Describe a means of generating and detecting PDM.

8-20-3   Describe a means of generating and detecting PPM.

8-21-4   Briefly describe four different methods for handling voice transmissions. Explain why PCM is strictly the only true digital system of the four.

8-22-4   Describe the meaning and importance of quantizing error in a PCM system. Why does a PCM TV transmission require more quantizing levels than a PCM voice transmission?

8-23-4   List the advantages of PCM systems. Explain why PCM is adaptable to systems requiring many repeaters and TDM. The information shown in Fig. 8-21 should be contained in this explanation.

8-24-5   Explain some possible applications for microprocessors in communication transceivers.

8-25-5   Why are telephone lines widely used for transmission of digital data? Explain the problems involved with their use.

8-26-5   Describe the function of each block of the computer data transmission system shown in Fig. 8-22. This description should include the functions performed by modems, UARTs, and line couplers.

8-27-5 Describe the difference between a half-duplex and full-duplex communication system.

8-28-5 Explain the meaning of "parity" as applied to a UART.

8-29-5 Explain the need for delay equalization when phone lines are used for signal transmission. If uncorrected, what would be the result of unequal delays to the different frequency components of a received signal?

8-30-6 What is a telemetry system? Explain why telemetry invariably involves multiplexing and pulse modulation techniques.

8-31-6 List the basic functions performed by a complete telemetry system and briefly explain them.

8-32-6 Why does the telemetry transmitting system shown in Fig. 8-27 have six subchannels feeding channel 4, while channels 1, 2, and 3 do not? Why is this arrangement better than using nine different channels, one for each condition being monitored?

CHAPTER 9

*9-1-2 Does the sound transmitter at a television broadcast station employ frequency or amplitude modulation? (4.253)

9-2-2 In what way is TV audio equivalent to broadcast FM radio, and in what way is it inferior?

*9-3-2 Does the video transmitter at a television broadcast station employ frequency or amplitude modulation? (4.252)

9-4-2 Explain the major benefit of combining AM and FM techniques in television broadcasting.

9-5-2 List and explain the functions of the six transducers used in a complete TV system.

*9-6-2 Why is a diplexer a necessary stage of most TV transmitters? (S4.80)

9-7-2 Describe the operation of a vidicon camera tube.

*9-8-2 What is a mosaic plate in a television camera? (4.256)

9-9-2 Sketch a resultant electrical video signal as would result from scanning the letter "E" in the setup shown in Fig. 9-3.

*9-10-2 In television broadcasting, what is the meaning of the term "aspect ratio"? (4.263)

*9-11-2 Numerically, what is the aspect ratio of a picture as transmitted by a television broadcast station? (4.270)

*9-12-3 What is the purpose of synchronizing pulses in a television broadcast signal? (4.257)

9-13-3 Provide an analogy between horizontal and vertical retrace as compared to reading a book.

*9-14-3 If the cathode-ray tube in a television receiver is replaced by a larger tube such that the size of the picture is changed from 6 by 8 inch to 16 by 12 inch, what change, if any, is made in the number of scanning lines per frame? (4.266)

*9-15-3 How many frames per second do television broadcast stations transmit? (4.264)

*9-16-3 Describe scanning as used by television broadcast stations. Describe the manner in which the scanning beam moves across the picture in the receiver. (4.255)

*9-17-3 Why is a scanning technique known as "interlacing" used in television broadcasting? (4.251)

*9-18-3 What are synchronizing pulses in a television broadcast and receiving system? (4.260)

*9-19-3 What are blanking pulses in a television broadcasting and receiving system? (4.261)

9-20-3 Calculate the frequency required for the horizontal sync pulses.

*9-21-3 What is the field frequency of a television broadcast transmitter? (4.268)

*9-22-3 In television broadcasting, why is the field frequency made equal to the frequency of the commercial (ac) power source? (4.265)

*9-23-3 Besides the camera signal, what other signals and pulses are included in a complete television broadcast signal? (4.259)

9-24-4 Describe the characteristics of a video amplifier.

9-25-4 Define resolution, vertical resolution, and horizontal resolution.

9-26-4 Explain why vertical resolution is less than the number (about 0.7) of horizontal lines. (*Hint:* Consider what might happen if a pattern of 495 alternate black-and-white horizontal lines were scanned by a TV camera such that each scan saw half of a white-and-black line.)

9-27-4 Calculate the horizontal resolution of a broadcast TV picture.

9-28-4 Calculate the decrease in horizontal resolution if the video signal bandwidth were reduced from 4 to 3.5 MHz.

9-29-4 Calculate the decrease in vertical resolution if the video signal bandwidth were reduced from 4 to 3.5 MHz, assuming that the horizontal resolution was not to change.

*9-30-5 How wide is a television broadcast channel? (4.274)

*9-31-5 If a television broadcast station transmits the video signals on channel 6 (82 to 88 MHz), what is the center frequency of the aural transmitter? (4.267)

*9-32-5 What is meant by 100% modulation of the aural transmitter at a television broadcast station? (4.277)

*9-33-5 What is the range of audio frequencies that the aural transmitter of a television broadcast station is required to be capable of transmitting? (4.276)

9-34-5   What TV channel is most likely to be heard on an FM broadcast receiver? Explain why.

*9-35-5   What is meant by vestigial-side-band transmission of a television broadcast station? (4.271)

9-36-6   Draw a TV receiver block, and briefly explain the function of each block.

9-37-7   State what a TV front end consists of and the important functions it performs.

9-38-7   List the three major types of VHF tuners, and briefly explain their characteristics.

9-39-7   Explain why a drift in dc supply voltage would cause detuning in a varactor tuner.

9-40-7   Explain the function of the switching diode referenced in the varactor tuner shown in Fig. 9-14.

9-41-7   Show how the VHF tuner is used in conjunction with VHF reception. Why is the VHF signal stepped down in frequency before it is given any amplification?

9-42-8   Calculate the sound and picture carrier frequency for channel 10 before and after frequency translation to the IF frequency (41 to 47 MHz). What is the required local oscillator frequency?

9-43-8   Explain the meaning of stagger tuning, and explain why it is often used in TV IF amplifiers.

9-44-8   Why is the sound carrier and its side bands only given one-tenth the amplification of the video by the IF response curve? Explain why part of the video signal is given less amplification also.

9-45-8   Discuss the function of a wavetrap and the need for such traps in TV receivers.

9-46-8   Explain why the sound IF will always stay at 4.5 MHz in intercarrier sets.

9-47-9   If an amplifier stage of the video section shown in Fig. 9-22 became inoperative, would the receiver's sound be affected?

9-48-9   In detail, explain the difference between adjustment of the brightness and contrast controls.

9-49-9   What is the function of dc restoration, and what kind of video sections require it?

9-50-9   With reference to Fig. 9-23, what is the function of a peaking coil?

9-51-10   What is the function of the sync separator? How is it able to differentiate between the horizontal and vertical sync pulses? What types of circuits are used for each?

9-52-10   Explain the operation of the vertical deflection system schematic shown in Fig. 9-25.

9-53-11   Explain the relationship between the horizontal deflection system and the

CRT anode high-voltage supply. Why is this a failure-prone area in a TV receiver?

*9-54-11   For what purpose is a sawtooth waveform used in a television broadcast receiver? (4.262)

9-55-11   What are the possible effects of a nonlinear deflection waveform?

9-56-11   Explain the operation of the horizontal system schematic shown in Fig. 9-28.

9-57-11   Explain the function of the damper system and flyback transformer.

9-58-12   Describe the process of interleaving, and explain its role in making color TV broadcast possible on the same bandwidth used in the monochrome system.

*9-59-12   Describe the scanning process employed in connection with color TV broadcast transmission. (4.288)

9-60-12   Describe the important features of the Y, I, and Q signals in a color TV broadcast.

9-61-12   Define the meaning of compatibility with respect to color and monochrome TV. How does a monochrome set properly display a color transmission?

*9-62-12   Describe the composition of the chrominance subcarrier used in the authorized system of color television. (4.285)

9-63-12   Explain how the Y, I, and Q signals are processed by a color TV receiver.

9-64-12   Why is extreme accuracy required of the color subcarrier oscillator within a color TV receiver? Explain how this accuracy is obtained in the receiver.

9-65-12   Explain the operation of the color killer. Describe the effect of a defective color killer.

9-66-12   Describe the important characteristics and construction of the color CRT. Include the need for convergence and how it is accomplished in this discussion.

9-67-13   In several paragraphs, describe a general troubleshooting procedure to be used for repair of a TV receiver.

9-68-13   List the probable defective stage(s) for the following symptoms:
(a) Video and raster normal, sound dead.
(b) Sound normal, video and raster dim.
(c) Raster normal, sound and video dead.
(d) Bent and "contrasty" picture.
(e) Floating picture, sound and raster normal.
(f) Loss of vertical sync.
(g) Loss of horizontal sync.
(h) Normal sound, no raster.
(i) No sound or raster.
(j) No color, black and white normal.
(k) Loss of one color.
(l) Loss of color sync.

INDEX

333